PRAISE FOR *THE HEALTHY BRAIN BOOK*

"An empowering guide for the care and feeding of your body's most precious organ."

**–David Perlmutter, MD, author of the *New York Times*
bestsellers *Grain Brain* and *Brain Wash***

"*The Healthy Brain Book* will be the gold standard for holistic doctors and the public for preventing and treating common brain ailments. I give it a 5-star rating for its cutting-edge content and easy-to-understand format."

–Mark Stengler, NMD, coauthor of *Prescription for Natural Cures*

"An engaging, encouraging, and empowering book with a focus on brain health—running the gamut from pediatrics to geriatrics—that makes it unique."

**–David L. Katz, MD, MPH, past president of the
American College of Lifestyle Medicine**

"*The Healthy Brain Book* is a gem, offering science, wisdom, and practical strategies that are essential to preserve and improve our brain power. Highly recommended!"

–Eva Selhub, MD, author of *Your Brain On Nature*

"*The Healthy Brain Book* is an outstanding addition to any library on human health. No one makes science more interesting, understandable, and relevant than Dr. William Sears. Packed with usable, easy-to-implement tips, it will answer every question you've ever had about how the brain works and why it matters for your overall health."

–Jonny Bowden, PhD, CNS, author of *The Great Cholesterol Myth*

"*The Healthy Brain* book is a complete and cutting-edge guide on how to take optimal care of what's between your ears. You can build your best brain at any age. In the most fun and interesting way possible, America's favourite pediatrician, Dr. Bill Sears, and world-renowned Alzheimer's expert, Dr Vincent Fortanasce, have teamed up to share their most recent, effective, and easy-to-understand tips on how to sleep, think, and remember better than you ever have before. It's time to go from neurotoxic to neuroplastic! If you want to have a smarter, sharper, happier, omega-3 rich, gluten-free noodle, just turn the page!"

–Bryce Wylde, DHMHS, guest expert and medical advisor for *The Dr. Oz Show*

THE
HEALTHY
BRAIN
BOOK

ALSO BY WILLIAM SEARS, MD

The Dr. Sears T5 Wellness Plan
The Omega-3 Effect
Prime-Time Health
The ADD Book
The Inflammation Solution
Sears Parenting Library titles

ALSO BY VINCENT M. FORTANASCE, MD

The Anti-Alzheimer's Prescription
End Back & Neck Pain
Life Lessons from Little League
Life Lessons from Soccer

ALSO BY HAYDEN SEARS, MA

The Healthy Pregnancy Journal

THE HEALTHY BRAIN BOOK

AN ALL-AGES GUIDE TO A
CALMER, HAPPIER, SHARPER YOU

William Sears, MD,
and Vincent M. Fortanasce, MD

WITH HAYDEN SEARS, MA

BenBella Books, Inc.
Dallas, TX

Illustrations by Deborah Maze

BenBella Books, Inc.
10440 N. Central Expressway, Suite 800
Dallas, TX 75231
www.benbellabooks.com
Send feedback to feedback@benbellabooks.com

BenBella is a federally registered trademark.

Printed in the United States of America
10 9 8 7 6 5 4 3 2 1

Library of Congress Control Number: 2019048628
ISBN 9781948836517 (print)
ISBN 9781948836760 (ebook)

Editing by Leah Wilson
Copyediting by James Fraleigh
Proofreading by Greg Teague and Dylan Julian
Indexing by WordCo Indexing Services, Inc.
Text design and composition by Aaron Edmiston
Cover design by Ty Nowicki
Printed by Lake Book Manufacturing

Distributed to the trade by Two Rivers Distribution, an Ingram brand
www.tworiversdistribution.com

Special discounts for bulk sales are available.
Please contact bulkorders@benbellabooks.com.

*To my wife, Martha, for helping to keep my
brain healthy for our fifty-three years of
marriage, and hopefully for many more.*

*To our eight children and fifteen grandchildren—may
you all enjoy a long and happy life with a healthy brain!*

–William Sears

*To my mother, Rose; my father, Michael;
and my wife, Gayl, who is my inspiration—always
positive, always giving, always present.*

–Vince Fortanasce

CONTENTS

CONTENTS

PART II: HOW TO USE THE SMART TOOLS TO HEAL 195

PART III: MORE QUESTIONS YOU MAY HAVE ABOUT BRAIN HEALTH   311

CONTENTS

AN OPENING NOTE FROM DRS. BILL AND VINCE AND COACH HAYDEN

Perhaps you are reading this book because you are already experiencing depression or anxiety, struggling with attention-deficit/hyperactivity disorder (ADHD), or feeling the early effects of Alzheimer's disease. This book will help you heal. Perhaps you are worried about developing one of these, and want to know the best way to prevent that from happening. We can help with that, too.

In our decades of running medical and health-coaching practices, and in our family lives, we have made brain health our research and our hobby. Not only that, all three of us have spent much of our professional and parenting lives managing "the Ds": depression, ADHD, obsessive-compulsive disorder (OCD), bipolar disorder (BPD), and AD (Alzheimer's disease). All of these "Ds" we simply call "stuff." Throughout this book you will read our personal journeys from debilitating "stuff" to vibrant brain health—through the use of our healthy brain plan.

In choosing what brainy health tips to give you, we first selected those brain-health tools from our decades of medical experience that worked for most people. Next, we chose those supported by solid science. Finally, we put all these tips and tools together into a healthy brain plan that will give you the fastest feel-good effects.

In Part I, you will learn how to acquire these top brain-health skills. Then, in Part II, we will guide you in using these self-help skills to heal from the brain imbalances you have, so that you may need to take fewer pills. Using our skills-before-pills model (and sometimes skills and pills), we will help you craft your *personal recipe* for better brain health. Last, in Part III, we'll answer many of the common brain-health questions you may still have.

BRAIN-HEALTH PROBLEMS—NOW THE TOP MEDICAL CONCERN

If you're worried about your brain health, you aren't alone. Consider these scary statistics:

- Neuroscientists predict that 40 to 50 percent of people will "lose their mind" by age eighty-five.
- Similarly, the American Academy of Neurology estimates that 40 percent of baby boomers will develop dementia at some point in their lives. When Dr. Vince started his neurology/psychiatry practice, he saw one patient a month with dementia. Now he sees three to four a week!
- For the first time in history, *cerebrovascular* disease is now more common than *cardiovascular* disease. The brain is now sicker than the heart.
- The American Academy of Neurology noted that 18 percent of the world's population—namely America and Europe—is responsible for nearly 50 percent of people with Alzheimer's.
- Insulin resistance and type II diabetes presently affect 60 percent of people nationally. In Dr. Vince's practice, 75 percent of the people he sees, especially those with dementia or early Alzheimer's, have insulin resistance, which is why Alzheimer's is sometimes called type III diabetes.

- By 2014, $84 billion had been spent on drug research and development to prevent, reverse, or stall Alzheimer's deterioration. All have failed.

Even if you may feel fine right now, the outlook is grim. So be smart! The best thing you can do for your brain health is to change unhealthful habits before you start feeling their effects. By the time you start feeling "sick and tired," mental and physical tissue damage is already done, and it takes much more mental and physical effort to reverse this damage than to prevent it in the first place. We call this *preventive medicine*—"preloading" yourself with a healthy brain and healthy brain abilities before you start experiencing disabilities. Whenever we are treating a patient with a debilitating brain illness, we are always thinking, "I wish I could have counseled this person years ago to preload their body and brain." If you start following our healthy brain plan now, your brain will thank you later.

KISMIF: KEEP IT SIMPLE, MAKE IT FUN

We wrote this book with your brain in mind. After insightfully listening to what readers want and studying how our brains like to read, we felt it was important for you to first understand *why* we wrote in the style you will read.

"Eye feel good." Since your brain is an extension of your eyes, if your eyes like what you see, your brain will also register "like!" The opposite is also true: if you clutter your eyes, you clutter your mind. In too many brain-health books, the print is too small, the paragraphs are too long, and the pages are simply not fun to look at. If your eyes feel good about what they see, your brain is more likely to enjoy and remember what you read. For this reason, we present each page with an "eye feel good" appearance. Enjoy your "eye candy"!

Seeing is remembering. The saying "a picture is worth a thousand words" is neurologically correct. Because the visual centers in your brain are so prominent, you enjoy and remember things more if they are partnered with pictures. At the end of many paragraphs we have an illustration that summarizes what you just read and visually imprints the message into your brain. *Read it. See it. Reflect* on it. Following this format, what you read about helping your brain health literally "sinks in."

PLAY SHOW AND TELL

Yes, a picture is worth a thousand words, but a *moving picture* is even better. Throughout this book, we give you links to watch what you just read. Simply enter the link into your phone's browser and tap "Go" to watch our most important brain-health tips, such as Dr. Vince doing isometric exercises, Coach Hayden doing yoga, and Dr. Bill narrating an award-winning animated version of how to open your personal internal-medicine pharmacy. Enjoy!

If you laugh, you will learn. In neuroscientist speak, this is called emotional arousal. When you smile, laugh, or even cry about a story or illustration, you are more likely not only to remember it, but to take action because of it.

That makes sense! Your brain loves logic. When something you hear or read makes you think "That makes sense!" it sticks in your memory and motivates the rational "I gotta do it" center in your brain.

To do this, we first rivet the reader with words that stick: "You're a *fathead*!"

Keep me young!

You're surprised, which turns up your alertness neurochemicals. You smile and wonder, "What on earth does 'fathead' mean? I have to know."

Then we follow with the explanation, "Your brain is mostly made up of fats," and the message, "Therefore, you need to eat more smart fats."

"Oh," you conclude, "that makes sense!"

Show me the science. You may want scientific studies to validate what your common sense tells you. We fulfill your wish with references in the Endnotes, page 349.

We relax your brain. Too many books exhaust the reader with information overload, causing what is known as "reader fatigue." Instead of presenting too much information, too fast, we help you catch the "core concept." Your brain likes to get right to the point. Each paragraph makes one point, and often opens with a short, bolded sentence or phrase telling you what you're about to read.

If you're still feeling overwhelmed by reading, try using the "thirty-second rule." If possible, after every thirty seconds of reading—the average time it takes to read an average paragraph—take five to ten seconds, look at the paragraph and the illustration, quiet your mind, and let the core concept sink in.

Are you now riveted to read?

> **Our main message: using this book as your guide,
> make brain health your hobby.**

We wish you happy brain health!

PART I

AN ALL-AGES, ALL-AILMENTS PLAN FOR BRAIN HEALTH

Perhaps you want to empower yourself with the smartest and most science-based self-help skills to smarter and happier living so you won't be one of those scary Alzheimer's statistics. These *preventive medicine* tools are for you. Or perhaps you or your loved ones already suffer from a brain-health problem, whether age related such as dementia, or any-age ailments such as depression, anxiety, or ADHD. These *healing* tools are for you.

We start in this section with a tour through your brain. The better you understand how your brain works, and the ways it is vulnerable, especially as you age, the more motivated you will be to help it behave smarter and happier. Then, we lay out the self-help skills of our healthy brain plan, which you can customize to best fit your needs, whether preventive, healing, or some combination of both.

In this guide to making brain health your hobby, we empower you by helping you:

- understand each "plant" in your brain garden and how to feed and fertilize it.
- eat smart foods, which we call "medicines for your mind."
- avoid dumb foods and drinks, which we call "anti-medicines for your mind."
- enjoy good gut health—your gut is your second brain!
- move more and sit less to live healthier and happier.
- think-change your brain by using smart stressbusters.
- enjoy sleep-well strategies for emotional health.

These tools stock your personal preventive medicine cabinet as you help yourself prevent and heal from depression, anxiety, Alzheimer's disease, and other causes of mental unwellnesses you will learn about in Part II.

Our wish for you: use more self-help skills so you need less pills.

Enjoy your trip through your brain and putting together your personal brain-health formula!

CHAPTER 1

~~~~~~~~~

GROWING YOUR BRAIN GARDEN

Imagine you're planting a garden. To get the garden to flourish, you have to plant healthy seeds; feed, fertilize, and water them; prune the plants; pick the weeds; and keep out pests. Likewise, to get your brain garden to flourish, you have to "plant" healthy brain cells, feed and fertilize them with *smart foods*, irrigate your brain tissue with *healthy blood vessels*, prune *unused brain tissue* (and replace it with new tissue), and keep out pesky *toxic thoughts*. You can do that!

OUR HEALTHY BRAIN PLAN AT A GLANCE

The healthy brain plan is designed to help you grow your brain garden. The plan has four different dimensions that you'll see come up again and again throughout the book.

To help you remember the four ways our healthy brain plan helps you grow your brain garden, we call it the LEAN approach to brain health:

- **L**—*Lifestyle*. Your lifestyle is how you live and learn, and how you use the tools of our healthy brain plan.
- **E**—*Exercise*. Science shows movers tend to have larger and healthier brains. In chapter four, you will learn how movement helps nourish your brain garden and increase the size of your brain.
- **A**—*Attitude*. Think happy thoughts! Happy thoughts grow the happy centers of your brain. Sad thoughts—we call them toxic thoughts, if you have too many and dwell on them for too long—grow your sad center. One of the most riveting reads in our whole healthy brain plan is chapter five, on how to grow your happy center and shrink your sad center. You will also learn how to keep toxic thoughts from polluting your brain garden.
- **N**—*Nutrition*. The brain, above all other organs, is affected by what you eat, for better and worse. In chapters two and three, you will learn what foods help keep all parts of the brain healthy. These supersmart foods act like smart fertilizer to help the brain garden grow smarter.

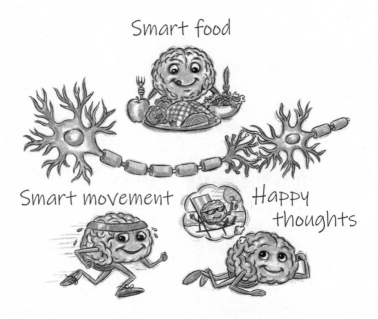

Smart food

Smart movement

Happy thoughts

TAKE A SMART AND FUN TRIP THROUGH YOUR BRAIN

Let's take a quick look at your brain's key "plants"—and how our healthy brain plan grows each one of them.

1. GROWING BETTER BRAIN CELLS

When you're growing a garden, you start by planting seeds. In your brain garden, the seeds are your brain cells: all 100 billion of them. And there are two parts of those brain cells that are key to making sure they are able to grow: the cell membrane, and mitochondria.

Mind your cell membranes. It's a medical truism that every organ of the body is only as healthy as each cell in it. And cell health begins with a healthy cell membrane. The cell membrane is like a flexible bag that holds all the genetic and energy-producing material inside your cell. When cell membranes are healthy, they are *selective* and *protective*. They let smart nutrients from the nearby blood vessels seep through into the cell and help it grow better, while screening out dangerous chemicals, called *neurotoxins*, that poison your brain garden. Think of a brain cell membrane as having millions of tiny entry doors, called *receptors*, that welcome nutrients into the cell and then send the waste material from energy generated within the cell back out, into the bloodstream. The healthier your cell membrane, the healthier your brain cells.

You're a "fathead." Cell membranes are composed mostly of fats. (Your brain tissue is 60 percent fats.) A healthy cell membrane needs the perfect balance of *flexible* fats (the omega fats you will learn about on page 32), which mold into whatever configuration the tissue needs, and *stiff* fats (healthy saturated fats), which keep the membrane from becoming too squirmy. An imbalance of these fats can cause cell membranes to

5

be too stiff, fragile, and leaky, so they are no longer selective and pro-tective. This means your cell membranes are highly influenced by what you eat.

BIRTH OF THE "FATHEAD"

Dr. Bill's longtime message to pregnant mothers, "You're grow-ing a little fathead!" began with his voracious appetite for learn-ing about why the human brain is mostly fat—one of the most meaningful messages he ever learned from his many "fat scien-tist" mentors.

Humans are the only species of land mammal that birth fat babies, and the theory is that extra adorable baby fat acts as a fuel insurance policy for the growing brain. As you will later learn, the smart fat diet that fuels the rapidly growing baby brain—momma's milk—highlights how important a smart-fat diet is, not only in those early years of rapid brain growth, but throughout life for brain maintenance and repair. (See chapter two for more about why eating healthy fats is smart for the brain—at all ages.)

It's your mitochondria, man! What hap-pens inside your cells is important for cell health, too. Your brain is called a hypermetabolizer, meaning it uses more energy per gram of tissue than any other organ in your body. What produces all this energy are thousands of microscopic energy generators called mitochondria, or "*mighty* chondria." The better you feed and care for these mighty mitochondria, the more energy they make. Mental fatigue, or "brain fog," is partially due to not eating enough smart foods to feed the mitochondria. There is even a "disease" in the doctor's dictionary to describe this—*mitochondrial dysfunction.*

Movement makes more mitochondria. Physical exercise is good for more than your muscle cells. It also builds more mitochondria in your brain, to make more energy for better brain performance. Besides *aerobics*, think *neurobics*.

2. BRANCHING OUT: FORMING YOUR MENTAL SOCIAL NETWORK

From each brain-cell seed in your brain garden extend branches—nerve fibers, called axons and dendrites—that make the brain cell look like a baby octopus or a tiny tree with many branches. Each brain cell may send out *ten to fifteen thousand* branches to other brain-cell branches. Like adding Facebook friends grows your social network, new branches grow your neuro-network. At birth, we have one hundred billion of these nerve fibers; by seven years old these nerves have grown trillions of branches. Imagine a little bush growing more "bushy" and connecting with other bushes throughout the brain.

Each of these branches is like an electrical wire along which nerve impulses travel. Simply put, how smart we are is influenced by how fast electrical messages travel down our nerve fibers and how many connections they make with other nerves.

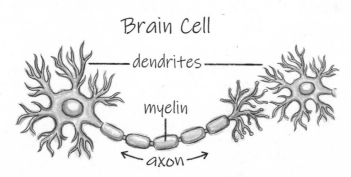

Make mighty myelin to make nerve impulses travel faster. The better you insulate the nerve fibers the faster and more efficiently electricity flows. The insulation on your nerve fibers is called *myelin*. Like insulation on an electrical cord, myelin keeps nerve impulses flowing along their intended pathways rather than leaking out. By increasing electrical efficiency, it enables nerve cells to carry electrical messages a hundred times faster than they could alone. So the more you can protect your myelin, the healthier—and smarter—your brain. Damaged or "frayed" myelin is the root cause of many brain illnesses.

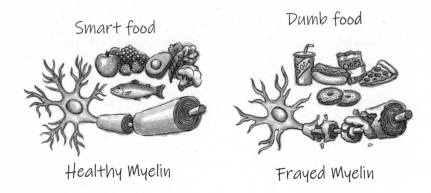

Smart food

Dumb food

Healthy Myelin

Frayed Myelin

Eat smarter foods, grow healthier myelin. Myelin is mostly fat. This is what makes it appear white—hence why it is called your "white matter," in contrast to the rest of your nerve tissue, or "gray matter." Because myelin is mostly fat, the "smarter" the fats we eat, the smarter we become. And because fat is prone to oxidation, or "turning rancid," we also need to eat foods that are high in antioxidants to protect our myelin. Our healthy brain plan helps you make smarter myelin.

My lesson from "Dr. Myelin." To learn more about how myelin gets made and gets frayed, we consulted the now late Dr. Bartzokis, Professor of Psychiatry at UCLA. He taught us about millions of myelin-making cells, called oligoden-drocytes (let's call them "O-cells" for short), which are spider-like cells that spin a web of myelin around each nerve, like bubble-wrapping a precious glass. The more wrapping, the more protection. When the myelin starts to wear out, like insulation on electrical wire fraying, these O-cells are programmed to repair and protect

Feed me smart foods.

nerve fibers by making more myelin. In fact, Dr. Bartzokis told us that he believed the ability to make more myelin is what made human brains more resilient than the brains of other animals. Because these O-cells work so hard—they use even more energy and produce even more exhaust than other brain cells—they need more supersmart foods to nourish and protect them. Neuroscientists call the nutrients in these smart foods *neuroprotectins*.

3. NOURISHING YOUR NEUROTRANSMITTERS: HOW BRAIN CELLS TALK TO EACH OTHER

How smart we are is also influenced by how well our nerve cells "talk to" neighboring nerve cells. There is a gap, called a *synapse* (Greek for "to bind together"), between cells in nerves and the brain. For nerves to communicate, biochemicals called *neurotransmitters* bridge this space like sparks flying across the gap. (So far more than sixty transmitters have been discovered, though we'll mostly be talking about four—serotonin, dopamine, norepinephrine, and GABA.) The end of a branch of one nerve cell acts like a sender, emitting these biochemical sparks that search for a receptor on the neighboring cell, like keys trying to fit a lock. As their name implies, neurotransmitters transmit biochemical text messages in your brain. The more quickly and easily the texts go back and forth between the brain cells, the better your brain works.

When neurotransmitters misfire. Smart and contented thinking depends greatly on how well these neurotransmitters help one cell talk to another. If these neurotransmitters misfire or misfit, your thinking misfires. Many mood disorders, such as depression, anxiety, and bipolar disorder, are thought to be caused by misfiring neurotransmitters. In neurospeak, there's a problem in the "gap" where nerve cells talk to each other. Treatment of these neurological and emotional issues relies upon getting the neurotransmitters to fire and fit properly. Smart foods help these neurotransmitters fire smartly. Think of it this way: fake food plugs the locks so the keys can't fit; real food helps the locks and keys fit. Since neurotransmitters are the language of your brain, you should nourish them.

NEUROTRANSMITTERS: THE NEUROCHEMICAL BASIS OF "YOU"

Your neurotransmitters are the neurochemical language or text messages that manage the thoughts and emotions that fill your brain. Their levels are what make you feel happy, sad, joyful, and all those fluctuating feelings you experience in daily living. The importance of "neurotransmitter balance"—the balance between all those individual neurotransmitters—is the current buzz among brain-health therapists. In chapter five, "Think Smart: Eight Stressbusters to Think-Change Your Brain," you will learn how to nourish your neurotransmitters into a happier "you."

4. MAKING HEALTHIER BLOOD VESSELS: IRRIGATE YOUR BRAIN GARDEN

Like any garden, your brain needs irrigation to deliver nutrients. Your brain is one of your most vascular organs. An organ is only as healthy as the blood vessels nourishing it, so it stands to reason that the organ that works the hardest enjoys the richest blood supply. Your brain uses *25 percent* of all the blood your heart pumps. Another reason your brain needs healthy vessels is to drain the excess heat and waste products produced by the brain's high metabolism.

> Think: The better my blood vessels, the better my brain.

When blood vessels fail. Blood vessel damage—more typically called a stroke—is one of the leading causes of disability and death. Some people demonstrate memory loss due to multiple small strokes, called *mini-strokes*. And dementia is aptly described as "hardening of the brain arteries." As you will learn in the following chapters, our healthy brain plan relaxes arterial walls, helps keep the vessel linings smooth, and the blood cells less sticky—all of which increase blood flow to nourish and protect your beautiful brain.

YOUR BRAIN-GARDEN PARTS

- *Brain cells*, or neurons, are the plants—all 100 billion of them.
- *Axons* and *dendrites* are like the limbs and branches of the brain cells.
- *Gray matter* is the neurons, both the brain cells and axons and dendrites.
- *White matter* is the tissue, such as myelin, and cells that hold the other brain tissue together.
- *Synapses* are the gaps where one brain cell connects with another.
- *Neurotransmitters* are the biochemical text messages that travel from one branch to another across the synapse.
- *Myelin* is the fatty insulation that helps nerves transmit messages faster.
- *Oligodendrocytes*, along with *astrocytes* and *glia cells*, help grow, maintain, and repair your neurons. They are the supporting cast that help the stars of your brain, your neurons, perform better.
- *Blood vessels* are the irrigation channels, supplying blood and nutrients to the brain.

WHERE IN YOUR BODY IS YOUR "MIND"?

For centuries neuroscientists have been searching the brain for the location of our "mind," as if there were a magical center for it. No one has ever found that center, most likely because they have been looking in the wrong places.

Beginning in the mid-eighties, neuroscientists started the-orizing—and later proving—that our "mind" is really the sum total of all the electrochemical and biochemical currents (what we call biochemical text messages) that travel throughout our brain and body, connecting the brain to all the cells in our body and to all of our other vital organs, especially our "second brain," the gut. With that insight, neuroscientists now use the more encompassing term "body-mind" to describe the part of us that manages our thoughts and feelings, instead of the half-truth "mind."

Throughout this book you will take many trips through your body-mind to connect the dots between how you think, feel, and act.

SIX SMART FACTS YOU MUST KNOW ABOUT YOUR BRAIN TO KEEP IT HEALTHY

Modern medicine, surgery, and medications enable us to live longer than at any time in human history. But even though our *lifespan* is longer, our *brainspan* (the years we live before losing some of our mind) is getting shorter. We may be living *longer*, but we are not living *smarter!*

In the first part of this chapter, you met the major plants in your brain garden. You will now learn how to nourish and protect them.

1. YOU NEED TO NOURISH YOUR "FATHEAD"

The smartest organ in your body is also the one most vulnerable to get-ting sick. Two biochemical features that make our brains so smart are also the ones that make it more vulnerable to disease.

We're "fatheads." Unlike tissue that is mostly protein, such as muscle, our brain tissue is 60 percent fat. As you just learned, that fat is crucial for providing myelin insulation on nerve fibers to make impulses travel faster. But that fat is also very vulnerable to attack from the number-one tissue damager, oxidative stress. Oxidative stress turns fat rancid. (Remember how that fatty fish fillet smelled after you mistakenly left it out all night? It oxidized.) But where does oxidative stress come from? Like exhaust produced from a hard-working engine, it comes from the brain itself.

We rust. As we've seen, the brain *uses more energy* than any other organ. While you may think that your heart would be the hardest-working organ in your body, it's really the brain. The heart is a relatively simple organ, automatically pumping blood 24/7, while the brain is amazingly complicated and continuously changes. Even though the adult brain weighs only three pounds, just 1 to 2 percent of your entire body weight, the brain uses 20 to 25 percent of all the energy from the foods you eat and 20 percent of the oxygen you breathe. All this "hypermetabolism" causes a lot of wear and tear, called *oxidative stress,* also known as rust. Aging brains in particular rust. Our healthy brain plan is really an anti-rust remedy.

Following the natural brain logic that is the hallmark feature of our book:

- "I'm a fathead, therefore I need to eat more smart fats."
- "I oxidize, therefore I need to eat more antioxidants."

That makes sense!

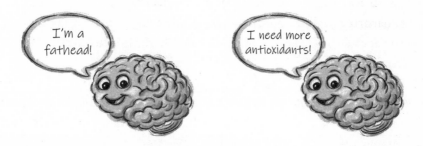

THE BRAIN IS YOUR FASTEST-GROWING ORGAN

In the womb, our brain begins as a single cell and multiplies to billions of cells by birth, with trillions of connections. The baby brain grows fastest in the third trimester of pregnancy, when it is making an estimated 250,000 connections a minute. The brain then triples in volume by two years of age and reaches 90 percent of its adult volume by age five. At age sixteen, the brain is full *size*, yet not fully *developed*. By twenty-four years of age, the brain prunes itself down to a mere five trillion connections and reaches a stage where we consider it to be "adult."

The brain continues to grow and transform throughout our life, especially the hippocampus, the center of memory and learning. Our brains are always forming new connections, and we can grow five thousand new brain cells each day by a process called neurogenesis. It's the growth of these new connections and cells that is essential to keeping our brains young.

2. YOUR BRAIN GARDEN NEEDS PROTECTION

As the main organ that makes us human, the brain should be the most protected. The good news is, besides being enclosed inside a thick skull, most of the tissues of the brain are defended by the *blood–brain barrier* (BBB), a one-cell-thick protective wrap that acts like a smart filter, letting in the nutrients the brain needs while screening out harmful chemicals that get into the bloodstream, called neurotoxins. The bad news is this barrier leaks. In fact, many neurodegenerative diseases can be attributed to a *leaky blood–brain barrier*. You've heard of "leaky gut." A related malady is "leaky brain." Our healthy brain plan protects your BBB from leaking.

A neurotoxin's trip across the BBB. The BBB is a picky eater, allowing smart foods through and excluding toxic foods—sometimes. Let's follow the artificial sweetener aspartame, a chemical additive that behaves artificially when it gets into the brain, on a trip through the body. Let's call it Toxi. (Dr. David Perlmutter, author of *Grain Brain*, and other noted neurologists teach that aspartame is a top "toxi.")

Toxi goes into the mouth where the taste buds sort of enjoy it, and that's where the fun ends. The first barrier it has to survive is the gut barrier, which screens out toxins that are bad for the body. Let's say Toxi gets through the checkpoint of the leaky gut and into the bloodstream. It makes its way throughout the rivers of the body and eventually gets into the rivers of the brain that irrigate the growing brain garden. If the BBB is healthy, Toxi will be kept out. If not, *Toxi forces his way through the leaks and becomes a neurotoxin.*

Vulnerability to "leaks." The BBB screening mechanism doesn't always work, especially at two vulnerable ages in the life of your brain: when you are a growing infant and young child, and when you are a senior. The younger and older the brain, the more vulnerable it is to having a leaky BBB. Because the rapidly growing brain of a child is more vulnerable to neurotoxins early in infancy, Dr. Bill's patients get his sermon: "Be a pure parent! Keep neurotoxins out as much as possible." Plus, if and how much the BBB starts leaking as we get older is very individual and depends on how you feed and protect it.

Another imperfection of the BBB is that it is not as protective in all parts of the brain, and some parts of the brain never develop a BBB at all. For example, the hypothalamus—the structure in the brain that most affects mood, memory, and learning—is very vulnerable, as is the pineal gland, the pea-sized structure that regulates melatonin, an important sleep hormone for brain health. The hypothalamus is also one of the areas of the brain that is most sensitive to neurotoxins, which may explain why memory and mood disorders top the list of brain-health problems, and are occurring at younger ages.

EYE FEEL SAFE

Since the eyes are really an extension of the brain, there is also a BBB in the vessels of the retina, the blood–retinal barrier (BRB). And we know much more about the blood–retinal barrier than the BBB, because eye vessels are easier to study than brain vessels. When your doctor shines a light through the pupil in your eyeball during a typical exam, they can actually see the tiny capillaries in the back of your eye.

Neuroscientists know that fruits, vegetables, and seafood, especially those that contain neuroprotectant bioflavonoids, such as astaxanthin and DHA, the smartest fat in seafood—key parts of the diet we'll introduce in chapter two—help keep the BRB from leaking. While we can't shine a light into the BBB of the brain like we can the eye, it seems safe to assume that foods that are good for the eyes are also good for the brain.

3. YOU CAN THINK-CHANGE YOUR BRAIN

You repair your brain. Doctors and drugs may help. Besides the protective effect of the BBB, brain tissue enjoys the healing property of *neuroplasticity,* the ability to repair damaged tissue, replace worn-out cells, and grow new and smarter brain cells. Researchers explain neuroplasticity as like an electrician working inside the brain, adding new wires and connections. Yes, you can teach an old brain new tricks!

People who are "disabled" in one part of the brain often become super-abled in another part. The adaptation of a blind person is perhaps one of the most vivid examples of how the brain is genetically programmed to make whatever changes are required to increase our chance of survival. In blind people, the area of the brain that is normally devoted to vision adapts to increase its sensitivity to touch, hearing, smell, and other senses. Dr. Bill remembers a blind mother bringing in her baby for him to check out the baby's rash. She could feel it; he could not see it. Curious and trusting Mom's instinct, Dr. Bill had her bring her baby back a day later, and sure enough there was a rash. She could feel the rash a day before he could see it.

A quick note regarding neuroplasticity: "Plastic" has a cold and hard connotation, but it comes from the Greek word *plastikos*, meaning "to mold." A baby's brain is the most moldable; second is the teen brain. But our brains are moldable at any age. We really shouldn't call people "hard headed"; that's not neuroscientifically correct. Nor is "rigid thinking." Brains are a work in progress!

You can think-change your brain health. You are not only what you eat. "You" are also what you think. One of the most exciting revelations of brain research in recent years is that our thoughts—those mysterious waves that we can't measure or see, but are obviously real—can affect, for better or worse, the neurochemicals our brain releases and the health of our brain cells. Our thoughts have the power to make our brains smarter and healthier. Conversely, our thoughts can dial up stressful neurochemicals that grow the *pessimist centers* of the brain. One of the major parts of our healthy brain plan is how you can think yourself happy, think yourself well, and think yourself smart.

SMARTNESS MEANS HAVING THE RIGHT CONNECTIONS

Even more important for "smartness" than the number of brain cells you have is the connections between them. Like branches from a tree trunk, nerve cells reach out, become more "bushy," and connect with neighboring bushy branches to smarten your brain garden. These connections are what allow us to do things automatically rather than having to think about every action; they let us do things more quickly and with more facility.

When you first begin to play the piano, for example, you have to deliberately look at each piano key and your fingers. With continued practice, more and more connections are produced in the brain—to the point that you no longer need to look at or think of the keys or your fingers. Instead, you play automatically. Unsurprisingly, concert pianists have twenty to fifty times the number of connections in those areas of the brain representing the fingers.

The younger you start playing, the easier it is to increase these connections, because the enzymes that form them are present in abundance in our earliest years to facilitate brain growth. The more exposure a child has to something when they're young, whether sports, arts, or intellectual pursuits, the greater their ability. This is the reason why early intervention on childhood disabilities, such as dyslexia, is so important.

4. YOUR BRAIN SHRINKS WITH AGE—IF YOU LET IT

The good news is the brain doesn't have to shrink as we get older. The bad news is that, in 40 percent of people over age sixty, it does. This is because what we call *brain growth fertilizer* (BGF), known neurologically as *brain-derived neurotropic growth factor*, the natural neurochemical that repairs old brain cells and grows new ones, tends to decline with age. But new research reveals that the brain doesn't have to get weaker as we get older, the way aging muscles do. We can grow new brain cells, repair worn-out ones, and even make lots of new connections between them *at any age* by increasing our BGF—something our brain plan helps you do.

How aging brains can become younger. *Problem:* Brain volume shrinks, the BBB leaks, nerve traffic slows, fatty tissue wears out, brain tissue rusts, blood flow slows, blood vessels shrink, stress wears out tissue, and tissue repair weakens. *Solution:* Exercise your brain, eat the supersmart foods we recommend (including lots of smart fats and antioxidants), and move your body to move your blood and increase BGF. Because natural neuroprotective mechanisms may weaken with age, the older we get, the cleaner we need to live, breathe, and eat, and the smarter we need to be about how food, thoughts, and lifestyle affect our brains.

Also, the "use it or lose it" analogy that applies to muscles also applies to the brain. An older couch potato is likely to lose a lot more bone and muscle. Yet new insights reveal that, as we get older, a person with a healthy diet and lifestyle does not necessarily have to lose bone or muscle—or brain tissue.

The moral of the aging story: the older we get, the more we have to work at keeping our brain smart and our body mobile. Fortunately, during seniorhood most of us have the luxury of *time* to do just that! When Dr. Bill hears, "You're getting older," he quips, "No, I'm entering my second childhood!"

THE AGING BRAIN: OLDER BUT WISER

Those with younger brains are quick to roll their eyes about those with older ones: "Grandpa forgot his car keys again" and "Grandma forgot my birthday." In some ways these jibes are justified. Older brains may be more forgetful and not as quick to master the latest technologies. In other ways the older brain is smarter, because the file cabinet of stored memories—*wisdom,* and how to react favorably to life's changing circumstances—is larger and richer. You may not be quicker, but you are wiser.

AGE WELL: HAVE A HAPPY HIPPO

Your hippocampus is your memory center, and taking care of it can help prevent the memory loss that comes with age. What makes your hippocampus so special, neuroscientifically speaking, is that it has a high "turnover rate," meaning it regenerates new brain cells at a rapid rate. That's good! But that also makes it vulnerable. Your hippocampus is a hypermetabolizer—it needs to be, to regenerate so many new brain cells!—meaning it generates a high level of *oxidation,* or wear and tear from the high metabolism. Therefore, our smart *antioxidant* eating plan is just what your hippocampus needs to stay happy.

5. YOUR BRAIN GETS STRESSED OUT

As we age, the natural stress-protective properties of the brain also lessen. Unresolved stress shrinks your brain. It literally wears you out. But don't get stressed by this news. The stress management part of our healthy brain plan, as you will learn in chapter five, can keep your brain from getting stressed out.

When you are overstressed, your brain actually gets "worried sick." Because brain tissue is so high in fat, it is easily damaged by stress hormones when their levels are too high for too long, a degenerating effect called *glucocorticoid neurotoxicity*. (Sounds like something you don't want to get!) Prolonged high levels of stress hormones have been found to "freeze" thinking and decision making. They damage connections in the brain and eventually lead to the death of brain cells, meaning your social network becomes smaller and begins communicating more slowly. The hormones that cause this are called *glucocorticoids*, the stress and aging hormones.

This is why stress management is one of our goals for brain health. Keeping your stress hormone levels in check can stop damage to the brain before it starts. As you will learn in chapter five, there are two stress management dials in your brain. One dials up a higher level of the stress hormones cortisol and adrenaline when you need quick energy and quick thinking, such as when you have to react fast to keep your kid from running into the street. Once the stressor is over, the second dial reduces the levels of stress hormones, like saying to the brain, "Cool it, danger over." Our healthy brain plan balances these stress neurochemicals in your brain to dial up when more are needed and dial down when the stressor is gone.

6. STICKY STUFF SLOWS DOWN YOUR BRAIN

Our simple yet scientifically correct explanation of the cause of most brain diseases is: *the more sticky stuff you put in your mouth, the more sticky stuff gets into your brain.* Sticky stuff goes by other medical names: oxidized cholesterol, inflammatory markers, C-reactive protein, homocysteine, high blood sugar level, hemoglobin A1c, and more. It's the kind of stuff your doctor sends you to a laboratory to have measured, because it can cause big trouble for your body and brain in two ways: causing neuroinflammation (which causes wear and tear) and slowing blood flow to vital organs. On the new food labels, sticky stuff goes by "grams of added sugar."

The sticky-stuff analogy provides a simple explanation for illness and aging: *The older you get, the more sticky stuff can accumulate.* For example, Alzheimer's, also called *cognitivitis,* is in part caused by *neuroinflammation* from the accumulation of sticky stuff.

Sticky stuff in the brain. In the brain, sticky stuff accumulates in two vital structures: the brain tissue itself, where it interferes with nerve transmission, and in the lining of the blood vessels, where it hinders blood flow. In fact, accumulation of sticky stuff in the lining of small vessels is one of the most common, yet preventable, causes of nearly all brain diseases, especially neurodegenerative diseases and strokes.

Keep the sticky stuff out!

In a nutshell, our brain-health prescription could be simply stated: keep sticky stuff out of your brain. In chapter two you will learn three important words to do just that: *avoid sugar spikes.*

A SMART HEALTH PLAN FOR ALL ORGANS

Because the body is only as healthy as the brain behind it, when you use our healthy brain plan you will enjoy what we call the *carryover effect:* the tools you learn to build a smarter, happier, calmer brain will also help all the other organs of your body be healthier, especially the following:

Eyes. The retina of your eyeball is an extension of your brain.

Gut. Appropriately called "the second brain" (see chapter three), your gut is closely connected to your brain by both large nerve pathways and biochemical text messages. Feel good in your brain, feel good in your gut.

Heart. Your brain controls your heart via a network of nerves that tell it how to behave: "Pump faster," "Pump harder," "Relax, man, you don't have to beat so fast and wear yourself out."

Lungs. Similarly, the brain is what prompts the lungs to "breathe faster," "breathe deeper," and "slow down."

Your body is a symphony orchestra where the hormones are the players—there are some 200 of them—and your brain is the master conductor. When the brain and the endocrine system are communicating well, you enjoy hormonal harmony, beautiful music, or what we call wellness. But if the master conductor, your brain, is out of sync with the players in your body, hormonal disharmony—illness—results. Teaching your brain to play beautiful music is the goal of this book.

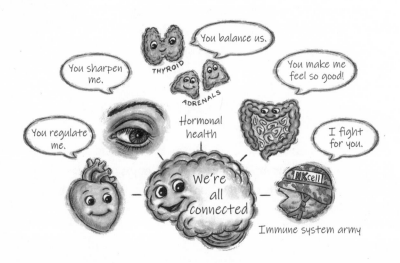

THE BROKEN BRAIN

Neuroscientists correctly label the times we live in as the era of "the broken brain." The causes of this modern malady are as follows:

1. *A shorter brainspan*. More seniors are losing their minds at an earlier age. This sad fact of mental living contributes to:
2. *Overwhelmed*. Adults caring for troubled kids and teens *and* aging senior parents have a double-stress whammy that wears them out and is worsened by:
3. *Information overload*. Being constantly hit with depressing news from media and through text alerts on our smart devices takes up too much cerebral real estate, leaving less room for joy. It not only saps mental energy, but also leads to us:
4. *Sitting too much—indoors—and eating too poorly*. As you will read throughout this book, these modern habits cause your brain garden to wither. We are also under:
5. *High pressure to succeed*. Many younger adults today feel too much pressure to get a prestigious job or hit other big life milestones, causing stress that takes its toll on the brain. Two terms that describe what we often hear from young adults are: *pressured* and *pessimistic* about their future.

These are the reasons why you need to start filling your personal toolbox now to prevent and fix a broken brain.

YOU HOLD THE KEYS TO BRAIN HEALTH IN THE PALM OF YOUR HAND

Now that you know that brain health is basically growing a smarter and healthier brain garden and enjoying the hormonal balance of happiness, here are the five "gardening tools" you will learn in the following chapters:

Eat Smart!

As we've seen, your brain is affected more than any other organ by what you eat, and eating smart foods is also the fastest way to change your brain. So, our healthy brain plan begins with chapter two on smart eating, with a special focus in chapter three on the gut–brain connection.

Move Smart!

In chapter four, get ready to learn about Nobel Prize–winning research on how movement smartens your brain, why movers feel happier than sitters, and the best movements and exercises for brain health.

Think Smart!

In chapter five, learn how to think-change your brain by managing those stress dials to grow your happy center and shrink your sad center.

Sleep Smart!

In chapter six, discover how, during quality sleep, your brain refreshes itself, repairs itself, and actually grows. Plus you'll learn tools to help you get a good night's sleep.

Here's one more key gardening tool to keep in mind:

Serve Smart!

When you share our healthy brain plan with others, you feel good by doing good. Serve your social network. Enjoying what we call "the helper's high" is a top feel-good medicine for your brain.

Enjoy your journey!

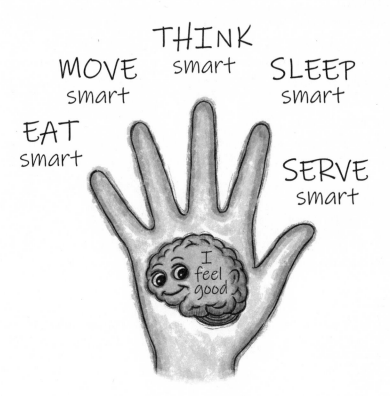

MOVE
smart

THINK
smart

SLEEP
smart

EAT
smart

SERVE
smart

I feel good

Your handful of health
and happiness

CHAPTER 2

~~~

EAT SUPERSMART FOODS: MEDICINES FOR YOUR MIND

S martly called "nutritional psychiatry," eating more supersmart foods, and fewer superdumb foods, is the fastest way to change your brain. What you eat directly affects the structure and function of your brain and, ultimately, your mood.

Here's why:

HOW FOOD AFFECTS YOUR BRAIN: AN OVERVIEW

What to Know	What to Eat
Your brain uses 20–25 percent of all the energy you get from food.	*Nutrient-dense* (we prefer "nutrient smart") foods that contain the most nutrients per calorie.

What to Know	What to Eat
Your brain is a *hypermetabolizer*, producing a lot of "exhaust" or *oxidants*.	Foods high in *antioxidants*.
Your brain is 60 percent fat.	A "smart fat" diet, not a low-fat diet.
Your brain uses 25 percent of all the *carbs* you eat.	*Smart carbs* that are slow-release and filled with fiber.
Your brain uses 25 percent of all the blood your heart pumps.	Foods that promote blood flow to the brain.
Your brain becomes sad, and functions badly, from sticky-stuff spikes.	Foods that naturally blunt sugar and sticky-stuff spikes.
Your brain is ultrasensitive to fake foods and chemical additives that can leak through (and weaken) the otherwise protective blood–brain barrier.	Real food that prevents leaky brain.

THE SAD DIET MAKES YOU SAD

Besides obviously being bad for your physical health, a poor diet is also bad for your mental health. Modern neuroimaging techniques, or windows into the brain, have proven what your mother proudly preached: "Junk food promotes junk learning, junk behavior, and junk feelings." Not surprisingly, scientific research has found that people who eat the SAD (standard American diet) are more sad—that is, they suffer more mood disorders. The SAD effect gets sadder. MRI scans have shown that people who eat the SAD diet tend to have a smaller hippocampus, the brain's memory center. The SAD diet of highly sweetened and highly processed fake foods triggers neuroinflammation, which literally shrinks your brain. To add more icing on this sad cake, research shows that depressed people tend to have higher blood levels of "sticky stuff," medically called "inflammatory markers." This is why neuroscientists in the know begin their brain-health treatment programs by prescribing a smart diet to quell the sad feelings from the SAD diet.

Why SAD is bad for your brain. The most recent explanation of the epidemic of mood disorders is that they are an inflammation of the mood centers of the brain, meaning these vital brain structures exhibit excess accumulation of sticky chemicals and show extra wear and tear as a result. This exciting new field of study is called *psychoneuroimmunology*—how the health of your immune system affects the health of your brain.

What you eat can either increase inflammation, by creating sticky stuff, or decrease it. After speaking at a brain-health conference in Hawaii, Dr. Bill was walking the beach with neuroscientist Paula Bickford, PhD, from the University of South Florida Center of Excellence for Aging and Brain Repair. The subject came up of how to simply explain how food improves brain health. Dr. Paula smartly responded, "I use the Tin Man character from *The Wizard of Oz* to explain oxidation of the brain. The Tin Man started to rust until Dorothy came along and oiled him."

EAT A SMARTER DIET

The four main features of our smart diet are:

1. It provides the most *nutrients* your brain needs.
2. It's rich in *antioxidants* to protect against oxidation.
3. It *prevents sugar spikes* that cause sticky stuff and damage sensitive brain tissue.
4. It *improves vascular health*. The better your cerebral blood flow, the smarter your brain garden grows.

EAT THE SMART SEVEN

In a nutshell, the foods you should eat for brain health are nutrient dense, high in *antioxidants*, and high in *smart fats* and *smart carbs*. They're "real foods"—meaning they are not processed and sugar-spiking fake foods, and they are free of or low in chemical additives.

The smart foods you're going to learn about are *nature's antidepressants* and brain-smartness builders. Follow this logic: the root cause of most mood disorders and other brain-health problems is *neuroinflammation* that lowers your happy hormones. Therefore, eat the foods that quell neuroinflammation. That makes sense!

1. SEAFOOD–THE SMARTEST FOOD WITH THE SMARTEST FATS

Suppose you consulted a top brain doctor and you begin your consultation with, "Doctor, I want to keep my brain sharp for many more decades and I'm a show-me-the-science person. Please give me the most science-based 'medicine,' which would also be the safest, that I can take to increase my chances of having a happier, smarter brain."

Top Doc smiles. "I congratulate you on wanting to take charge of your brain health." And she quickly scribbles on her prescription pad, "Go fish!"

In deciding what merited the title of top superfood, we used common sense: the closer the nutrient profile of the food was to that of the brain, the smarter the match. Your brain is mostly fats. Seafood is mostly fats—smart ones. Therefore, seafood is a smart brain-health food. Make sense?

Wild salmon wins our top supersmart food award. Dr. Bill's top brain-food choice is wild Pacific *king* salmon. There is no other superfood on the planet that contains more nutrients your brain needs. Look at the average nutrient profile of a 6-ounce fillet (280 calories) of wild Pacific salmon, and see how it matches your brain's nutrient needs:

Smart Nutrients Seafood Provides	% of Daily Recommendation	Why Your Brain Needs It
Omega-3 DHA, 1,200 milligrams Omega-3 EPA, 900 milligrams	200%*	The top fats brain tissue needs (see why, page 32).
Protein,** 40 grams	40–50%	Builds new brain tissue.
Astaxanthin,*** 8 milligrams	100%	One of nature's most powerful antioxidants and what makes salmon pink.
Vitamin B_{12}, 9 micrograms	160%	B12 deficiency can predispose the brain to dementia.
Vitamin B_6, 2 milligrams	60%	Promotes healthy brain tissue.
Selenium, 75 micrograms	80%	A powerful antioxidant against neuroinflammation.
Vitamin D, 2,000 IU	100%	Powerful antioxidant.
Choline, 190 milligrams****	40%	Promotes brain cell maintenance and growth, and facilitates neurotransmitter functions.
Niacin, 16 milligrams	83%	Protects against neurodegenerative diseases.
Iodine, 100 micrograms	100%	Promotes thyroid health.
Tryptophan	100%	Promotes good sleep, which promotes emotional health, and makes serotonin.

Smart Nutrients Seafood Provides	% of Daily Recommendation	Why Your Brain Needs It
Carbs, 0		Doesn't cause sugar spikes!

*While the official USDA daily value (DV) for these fats is not yet available, top seafood scientists recommend for adults at least 1,000 milligrams a day of combined DHA and EPA per 2,000 calories consumed.

**For optimal brain health, we advise consuming an average of 1 gram of protein per pound of body weight, and slightly more for professional athletes.

***While the official USDA DV is not yet available, we generally recommend 8 milligrams a day, perhaps more for professional athletes (see more about astaxanthin, page 62).

****While there is no official USDA DV, neuroscientists suggest 500 milligrams a day.

See page 35, for a safe seafood source we recommend!

A SMART SEAFOOD MATCH

Omega-3 fats, the smartest fats, make up around *12 percent* of the fats in the brain and around *12 percent* of the fats in wild salmon.

VEGAN SEAFOOD

You can be vegan and still eat smart seafood! Although the *top* smart seafood is wild salmon, we use the term "seafood" here to include both fish and *seaplants*, which contain *algae,* a rich plant source (and therefore a *vegan* source) of omega-3s.

Why can't vegans just get their omega-3 fats from land plant sources, such as seeds, greens, and oils? ALA (alpha-linolenic acid), the most abundant omega-3 fat in nature, is what we call a "short guy"—meaning it has a chain length of eighteen carbon atoms. Yet the brain prefers to grow, maintain, and repair itself using mostly omega "tall guys"—EPA (twenty carbon atoms) and DHA (twenty-two carbon atoms).

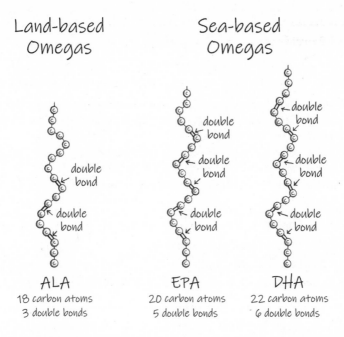

Land-based
Omegas

Sea-based
Omegas

double
bond

double
bond

double
bond

double
bond

double
bond

double
bond

double
bond

double
bond

double
bond

ALA
18 carbon atoms
3 double bonds

EPA
20 carbon atoms
5 double bonds

DHA
22 carbon atoms
6 double bonds

The double bonds act like hinges, which make the fat flexible and able to mold itself into the brain cell membranes, enabling better neuroplasticity. Because they have more double bonds, sea-based omegas (EPA and DHA) are more flexible and therefore the smartest fats in your brain.

The body is technically able to turn short guys into tall guys. The liver makes an enzyme, appropriately called "elongase," to tack on two to four more carbon atoms to the short guys, ALA, to convert them into the tall guys, EPA and DHA, that the brain prefers. Biochemically this is known as omega-3 conversion. Yet people vary in how well their individual biochemistry can convert ALA to EPA and DHA. You may be a low converter or a high converter. In fact, most people only convert between 2 and 4 percent of the omega-3 ALA oils they eat into the tall guys.

In our medical practices, when we see a person who prefers to be vegan, we measure their omega fat levels (see "Measure Your Omegas," page 45) to make sure they're getting the omega-3s they need. One of Dr. Bill's patients, a nurse and confirmed vegan who knew about the importance of omega-3s in her diet, insisted she was getting enough omega-3s from her daily tablespoon of ground flaxseeds, but when we measured her omega-3 red blood cell index, it was only 3.6 percent, well below the optimal level of 8 percent. Being a wise nurse, she agreed to take a daily supplement of omega-3 from a seaplant source (algae).

DR. BILL'S FAVORITE FISH STORIES

In my quest to learn about smart fats, I spent twenty years speaking and working with top "fat scientists."

My medicine meal. I experienced a vivid demonstration of the brain-boosting effects of seafood while on a speaking tour of Japan in 2001. Before a talk, our Japanese hosts took my wife, Martha, and me out for a "medicine meal," a century-old tradition to serve a visiting professor a special meal before giving a lecture. At lunch we were presented with twelve little strange-looking dishes on a tray. Our interpreter translated the dishes as "edible seafoods," a combination of many different fish and sea plants. Within an hour of eating, my brain was amazingly alert. I also enjoyed "good gut feelings," neither hungry nor full, just perfectly satisfied. Those good feelings stayed with me for at least four hours. It was one of the best talks I've ever given!

My sticky stuff story. In June 2013, I was invited to present omega-3 science on brain health at a GOED (Global Organization for EPA and DHA) conference in Tokyo, Japan. In my attempt to make the science simple and fun, I used an analogy of a riverbank. When you eat a diet high in sticky stuff (processed food) and low in seafood (smooth, non-sticky food), the sticky stuff builds up on the lining of your blood vessels like garbage building up on the edges of a riverbank. Arterial health is about having a clean riverbank. I noticed the interpreter saying "sera sera" (clean river) and "doro doro" (dirty river) and the audience nodding yes and no. After the talk, many healthcare professionals came up to me saying "sera, sera" while bowing approvingly, and shaking their heads to "doro, doro." The analogy, shall we say, stuck.

Survival of the fattest. That evening, Hayden and I had the privilege of dining with Professor Michael Crawford, director of the Institute of Brain Chemistry in Human Nutrition, London, and other "fat scientists." As we ate our omega-3-rich meal, Dr. Crawford gave us his theory as to how human brains got bigger and humans got smarter: when humans migrated from central to coastal areas and began eating more seafood, the human brain tripled in weight, from one pound to three pounds. My take-home message: *humans got smarter as our brains got fatter.*

Why salmon are pink. While writing *The Omega-3 Effect*, I went fishing for more seafood science with my favorite fisherman and my most trusted seafood source, Randy Hartnell, owner of Vital Choice Seafood. While watching salmon swim and jump their way upstream, I asked Randy, "Why are salmon pink?" His "wow!" answer: during its vigorous swim upstream, a salmon's body and brain work so hard that these tissues suffer lots of *oxidative stress*. Without extra protection, they'd rapidly wear out. Enter Dr. Mother Nature, who genetically infuses into the hardworking salmon nature's most pink and powerful *antioxidant*, called *astaxanthin*. (See more about astaxanthin, page 62.)

Dr. O. Mega III Boasts, "I'm The Smartest Fat"

Imagine a "help wanted" ad from a brain-health agency: "Wanted: Experienced expert on the 'Ds': ADD [attention deficit disorder], ADHD, BPD, OCD, CVD [cardiovascular disease], AD, and depression."

The world's top expert, we'll call her Dr. O. Mega III, applies and explains why she's perfect for the job: "I'm very credentialed; there are thousands of scientific articles to support that I'm the smartest nutrient. I help every one of the Ds you mentioned."

The company CEO challenges her: "Why do you claim to be the smartest nutrient?"

I'm a neurologist, ophthalmologist, cardiologist, gastroenterologist, psychologist and a psychiatrist.

Dr. O. responds, "In the scientific community I'm known as the 'flexible fat,' meaning I can do any job you need me to do. I can grow and repair brain tissue. I can make brain tissue work smarter. I'm a smart fit for nearly all brain needs at all ages."

"That makes sense," the CEO replies. "You get the job! I will send out a memo telling our organization to eat more of you."

Remember the tag "flexible fat." The root cause of most illnesses, especially in the highly vascular organs and in fatty tissue like the brain, is that tissues get stiff. Stiff tissues don't work as well as flexible ones.

Meet the "fat boys." While writing this book, Dr. Bill hosted a two-day roundtable at his home attended by six top scientists who have collectively published more than 1,800 scientific articles. During his meetings with the "fat boys" and "fat ladies"—as Martha calls them—he learned that omega-3 fats help make four important structures of the brain healthier: brain cell membranes, myelin insulation, nerve connections, and cerebral blood vessels.

Omega-3s make brain cell membranes healthier. As you learned on page 5, cellular health leads to organ health, and cell membranes protect the health of the whole cell. The omega-3 fats DHA and EPA make brain cell membranes more flexible, protective, and selective, letting in healthy nutrients and keeping out neurotoxins.

Omega-3s make healthier myelin. Omegas act like *neuroprotectants* to protect vulnerable myelin against oxidation or, in fat language, turning rancid. Myelin damage can even be reversed by eating more omega-3 fat. And many neurodegenerative diseases, such as Alzheimer's and Parkinson's, are thought to be diseases caused by damaged myelin. Like frayed electrical wires, malfunctioning myelin causes thoughts to become slower and foggy, leading to forgetfulness, and muscle movements to become less coordinated. Be mindful of the correlation between the food you eat and the myelin you make.

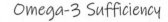

Omega-3 Insufficiency   Omega-3 Sufficiency

Frayed Myelin   Smart Myelin

You may have heard the term "healthspan," which means the number of years Americans tend to enjoy good health. While our "lifespan" is getting longer, averaging around *77 years*, our healthspan is not: it now averages around *43 years*. How about our "brainspan," the number of years before we begin "losing our mind"? Although brainspan is highly individual, myelin production seems to peak around age 45 and thereafter declines—unless we use a healthy brain plan to nourish and protect it.

Omega-3s feed neurotransmitter activity. Remember Brain Function 101: The more connections you make between brain cells and the faster chemical text messages are sent to and received by other cells, the more intelligently you think. Think of these messages as high-speed ferryboats traveling across the synapse between sending and receiving brain cells. The ferryboat needs to dock at the right receptor to transmit the information it carries. Omega-3s feed both the ferryboats (messages) and the docks (receptors). They act like sculptors to ensure the receptor sites welcome and exactly fit incoming ferryboats. So if you don't eat enough omega-3 fats, the messages, or ferryboats, don't travel fast enough, and once they get to their destination, they can't unload their messages because they can't fit into the docks.

Omega-3s keep brain vessels less sticky. Of all nutrients on the planet, omega-3s are the best at keeping sticky stuff out of blood vessels. They work their way into the membranes of red blood cells and the vessels themselves, acting as a lubricant to keep the blood cells from sticking together or to vessel walls, which makes blood flow smoother and faster.

Protecting your microvasculature (the tiny blood vessels in your brain) is one of the most therapeutic brain-health changes you can make, and that's exactly what omega-3s do. In fact, that's how their health effects were first discovered in the early 1970s. By keeping sticky stuff off the lining of the blood vessels, where many natural brain-health medicines are made (we'll learn more about that in chapter five!), they increase the release of two particular natural medicines that your body makes to protect your microvasculature: *protectins* and *resolvins*.

Omegas make blood flow better.

Having just the right amount of omega-3s in your blood vessels is like having traffic directors on a highway during rush hour, making traffic flow smoothly and less congested. And the more blood flows to your brain, the healthier it is.

DR. BILL'S HEART-SMART FISHING TRIP

I spent a week in Norway fishing—and learning—from Dr. Jorn Dyerberg, deservedly called the father of omega-3 science. Like many doctors in the early '70s, Dr. Dyerberg questioned the low-fat advice becoming popular then, mainly because of a statistic that set the low-fat world on its behind: people who eat the most fats, such as the Inuits in Greenland (more than 50 percent of their calories come from fat), have the lowest incidence of heart disease.

After much research, Dr. Dyerberg and other "fat experts" concluded that it wasn't the amount of fats that caused heart disease—and brain disease. It was the type of fats. Good fats protect against brain and heart disease; bad fats don't.

EATING SEAFOOD TO PREVENT DEGENERATIVE DISEASES

The more seafood you eat, the less likely you are to get neurodegenerative diseases, not just because of the anti-inflammatory and anti-sticky-stuff effect of omega-3s, but also because of the other antioxidants in seafood, such as astaxanthin, selenium, vitamin D_3, and the others listed on page 31.

Now that you understand how omega-3s help the structural health of the brain, you may wonder: "How exactly does that make my brain happier and smarter?"

Research has shown that people with adequate levels of omega-3s:

Healthier cells
Faster communication
Better blood flow
Less neuroinflammation

- have larger brains;
- have smarter brains;
- suffer less depression and anxiety;
- suffer less Alzheimer's disease;
- suffer less neurodegenerative diseases, such as Parkinson's and multiple sclerosis; and
- show less behavior and learning Ds: ADD and ADHD.

"When you change your fats, you change your brain," Captain Joseph Hibbeln, MD, the leading omega-3 researcher in the U.S. military, told me during a dinner with other omega-3 scientists.

Buddies in brain health

Seafood Is Mood Food

When we surveyed scientific studies and attended conferences on nutrition and brain health, one of the top nutrients that kept coming up in connection with mood disorders, especially for depression and bipolar disorder, was omega-3s. This makes sense.

Meet the "merry omegas." If mood disorders result from misfiring or mis-wiring of neurotransmitters in the mood centers of the brain, and omega-3s EPA and DHA are the main structural and functional molecules of nerve firing and wiring, then feeding your brain enough of these smart fats is key to maintaining a balanced mood.

Throughout this book we try to put good science and good sense together. If mood disorders are a biochemical imbalance in the brain's messengers—neurotransmitters such as serotonin, dopamine, norepinephrine, and GABA—and omega-3s feed these neurotransmitters, it stands to reason that omega-3s would help patients with mood disorders. In Dr. Bill's medical practice, they dub omega-3s "happy medicines" or "merry omegas" because they help the brain make more happy hormones.

In a nutshell, these "merry omegas" are mood mellowers: they help raise the "lows" (depression), lower the "highs" (manic feelings), and put the brakes on the impulses of people with OCD. As omega scientist Dr. Bill Lands, author of *Fish, Omega 3s, and Human Health,* told Dr. Bill while dining on salmon at sunset at his and Martha's home: "The sea may give us food that lessens our overreactions to the stresses in our lives on land."

More omega-3s, less stress. Feeling "foggy" has a neurochemical basis. When brain cells are constantly bombarded with excess stress hormones from chronic, unresolved stress (called glucocorticoid neurotoxicity, as you may recall), the brain cells eventually wear out. Omega-3s help protect brain tissue, particularly the vulnerable connections between brain cells, from attack and damage by excess stress hormones. And some areas of the brain, especially the *hippocampus* and *hypothalamus* (your mood centers), are more sensitive to and needy of smart nutrients than others.

Eat more seafood, shrink less brain. More about how science and common sense agree: PET scans show that severely depressed people suffer from shrinkage of their hippocampus, another area of the brain that is made stronger by omega-3s. Besides boosting the happy hormones in the happy centers of the brain, the omega-3 EPA (one of the top omega-3 fats in seafood) helps smarten and heal these depressed brain centers by:

- *improving the blood flow* to these areas;
- *decreasing neuroinflammation,* the root cause of many brain illnesses; and
- *increasing* BGF, your natural brain garden fertilizer.

SMART FATS GO TO SCHOOL

During another salmon-at-sunset dinner at our home, Cornell University neurology researcher Dr. Tom Brenna told me, "Kids who eat more omega-3s learn better." (See "Endnotes" for amazing studies on how eating sufficient omega-3s help students learn and behave better.)

Omegas help me do better in school.

Omega-3s alleviate Alzheimer's. You may have read the bad news that no prescription drugs have been shown to reverse, delay, or prevent Alzheimer's. The good news is there is a different kind of "medicine" that science shows can help prevent and delay the progression of this dreaded disease—one that tastes good and has no harmful effects. That's right: omega-3s!

- Brain tissue samples from patients with Alzheimer's show low levels of omega-3 DHA.
- Some studies show that the more omega-3s people eat, the less their risk of dementia.
- The Zutphen Elderly Study showed that people who ate more omega-3s showed slower cognitive decline than people who had an omega-3 deficiency.

Alzheimer's, like its fellow neurodegenerative diseases such as Parkinson's and multiple sclerosis, is an inflammatory disease, so it makes sense that a powerful anti-inflammatory, like omega-3s, can help.

SEAFOOD IS *SEE*FOOD

What's smart for the brain is also healthy for many other systems in your body. When Dr. Bill began studying why seafood is smart food and eating more of it, he noticed that his vision was improving. One night while attending a baseball game, he realized he had forgotten his glasses. He looked at the distant scoreboard and could read it very clearly. He wondered: Why was his vision so much better? Then he remembered reading a pediatric medical journal that showed that children whose mothers ate more seafood during their pregnancies tended to have better vision. Could it be that seafood is *see*food?

That makes sense for several reasons. Retinal tissue is primarily fat and blood vessels, and omega-3s exert their best protective effects on fatty tissue and blood vessels. The retina also has the highest metabolic rate and oxidative stress (wear and tear) of any tissue of the body, and seafood contains three potent antioxidants in addition to omega-3s: selenium, vitamin D, and astaxanthin. Lastly, photoreceptor cells in the retina are so high functioning that they wear out quickly. They need to be replaced almost daily by new cells, which require a continuous flow of omega-3s.

—Retina

Good science and good sense go together: Seniors who eat more omega-3s have a lower incidence of the leading cause of blindness, ARMD (age-related macular degeneration). People who eat more omega-3s have healthier retinas. Go omega-3s!

"But I Don't Like Fish!"

When we give our patients the prescription "Go fish!," some tell us they don't like the taste. In return, we ask them to try what we call "the taste-transformation test." You can change the programming in your brain that makes you believe "I won't like this."

Try this exercise: Put a small fillet of wild Pacific salmon (which you don't like) on a plate surrounded by foods that you do like. Put your favorite sauce on top of the salmon. Before you take your first bite, imagine all the nutrients in that bite of salmon helping your brain work smarter, your eyes see better, your heart pump stronger, your skin look and feel better, and your gut feel better. Imagine a head-to-toe health makeover. If you begin each bite with this intellectual process, focusing on "it's so good for me," over the course of ten to fifteen tries you may go from "I must learn to like it" to "I like it a little" to "It's okay" to "I like it a lot!" You will most likely not only develop a taste for seafood, but eventually *crave* it.

If you try this transformation test and still don't like fish, there are also omega-3 supplements you can rely on. (See page 62.)

OUR TASTY FISH TIP

Try your favorite toppings. Ours is a honey glaze with a mixture of Dijon mustard, lemon sauce, and dill, sprinkled with capers. Pairing seafood with wild rice is yummy!

How Much Seafood To Eat And/Or What Omega Supplements To Take For The Brain Health You Want

When it comes to brain foods and brain supplements, one dose doesn't fit all. Why? Because of a nutrient feature called *bioavailability*, or how available certain nutrients are to the body. In other words, it's not how much of a food you eat that counts, but how much of that food's nutrients your intestines *absorb*, and that can be as individual as our personalities. You may be a high absorber, a medium absorber, or a low absorber. The good news is that new blood tests can provide an answer to whether you're getting enough omega-3s, so that you can adjust your intake accordingly.

MEASURE YOUR OMEGAS

Before you start with our recommended dosages, use a simple fingerstick blood test to measure the percentage of the omega fats in your red blood cell membranes. This won't tell you the exact percentage in your brain, but because we can't take a piece of your brain and measure it, measuring the concentration in your red blood cell membranes is a good indicator. We recommend the *Vital Omega-3/6 HUFA Test* from www.VitalChoice.com. (See "Endnotes" for how we use this test in our medical practices.)

Of all the fat values you will get back on the printed sheet (there will be many), the two to pay the most attention to are:

- The *red blood cell omega-3 index*: the percentage of omega-3s in your red blood cell membranes. You want this number to be *above 8 percent*.
- Your *omega balance*: the ratio of omega-6 to omega-3 fats in your red blood cell membranes. You want no higher than 3:1 omega-6:omega-3; a 2:1 ratio is even better (see "How to Balance Your Omegas," page 77).

If you have these healthy omega numbers, then the amount you absorb from what you eat is very good. If not, the next section includes our recommendations for eating more seafood or taking a higher dose of omega-3 supplements.

Dosage recommendations. The general guideline recommended by top scientist groups, such as the International Society for the Study of Fatty Acids and Lipids, the American Heart Association, and the National Institutes of Health, is 1,000 milligrams (1 gram) a day of omega-3s per

2,000 calories consumed. That 1,000 milligrams should include a roughly 1:1 ratio of DHA to EPA.

A gram a day, I say!

We recommend that adults who are fond of fish should eat at least *two fistfuls* (about 12 ounces) of wild seafood, such as wild Pacific salmon or line-caught tuna, per week. This will give you about 500 milligrams of omega-3 fats a day, or about half of the minimum amount we recommend for brain health. Add to this a daily supplement of *500 milligrams* DHA/EPA, and you have 1,000 milligrams of omega-3 DHA/EPA per day.

While 1,000 milligrams of omega-3 fats is a healthy amount for a person who tends to eat 2,000 calories per day, a very active person, such as an athlete who eats 3,000 calories a day, would need about 1,500 milligrams of omega-3 fats daily.

If you have cardiovascular disease or a neurological disorder, such as depression, bipolar disorder, or neurodegenerative diseases, your healthcare provider may advise 2,000 to 3,000 milligrams per day.

Smart Omegas

Each serving contains:
DHA 500 mgs
EPA 500 mgs

It's difficult to reach a level of 2,000 to 3,000 milligrams a day by only eating seafood, so in the case of neurological or cardiovascular disease, we advise taking an omega-3 supplement. (See AskDrSears.com/omega supplements for the sources we recommend and any updates on recommended dosages for particular illnesses.) (See chapters eight and nine for how omega-3s can help prevent Alzheimer's disease and ADHD.)

~~~~~~~~~~

Seafood is the smartest fat with the most science on its health benefits. Yet brains do not get smarter on seafood alone. We've listed the next six smartest foods in the order of priority, based on two criteria: how high their *nutrient density* is (i.e., which ones provide the most nutrients your brain needs per calorie you eat), and how much scientific research there is showing their brain-health effects.

2. EGGS ARE EXCELLENT

Once upon a time, eggs were shunned because they are high in cholesterol and half of their calories come from fat. But your brain is also high in cholesterol, and more than half of your brain is fat. Plus, at a measly 75 calories apiece, eggs are dense in nutrients that are good for your brain.

Like the "fat is bad, sugar is good" debacle, where faulty science didn't match up to common sense, the idea that eggs are bad for you was finally exposed as false. Recent insights have exonerated the egg from contributing to heart disease, and even cardiologists are now promoting eggs as a heart-healthy food. For most people, there is little association between the cholesterol in the foods you eat, the cholesterol in your blood levels, and its relationship to heart disease. (There are, however, a small percentage of people whose blood cholesterol may go up when they eat more cholesterol.)

SCIENCE SAYS: CHOLESTEROL IS A SMART NUTRIENT

Not only is cholesterol not bad for the heart, it's also good for the brain. Dr. Bill's interest in cholesterol began with the discovery that the rapidly growing baby brain has a voracious appetite for more cholesterol. It makes sense! Cholesterol is a top structural component of brain cell membranes and myelin—that smart fatty insulation that makes nerve traffic travel faster. As an insurance policy against not eating enough cholesterol, the brain is even able to make its own! These fat facts are why more and more doctors are becoming increasingly selective in prescribing drugs that mess with the body's cholesterol production.

What makes eggs so smart? Both the white *and the yolk*. Next to the proteins in mothers' milk, the protein in egg yolk wins the "Top Protein" award, as judged by its "biological value," meaning how efficiently the body utilizes the protein.

"I'll have an egg-white-only omelet, please." Not smart! Egg yolk contains as many, if not more, brain-healthy nutrients than the white. The yolk not only increases HDL, the "good" cholesterol, but it also makes the particle size of LDL bigger. This larger particle size helps keep LDL from sticking to the sides of your blood vessels and worming its way through the lining into the arterial walls, where it can lead to buildup of plaque and stiffer arteries. Small-particle LDL is the "bad cholesterol." This is why measuring "LDL particle size" is an important addition to the standard cholesterol profile blood test. Remember, as you learned in chapter one, what's good for your vessels is also good for your brain.

Egg yolk is also the highest source of the nutrient *choline* (the second highest source is seafood), which helps build healthier brain cell membranes. Also, choline builds *acetylcholine*, one of the neurotransmitters that sends biochemical text messages between brain cells and from your brain to your body.

Lastly, egg yolk contains *lutein and zeaxanthin*, carotenoids that help protect and maintain healthy brain (and eye) tissue by quelling neuroinflammation.

The best eggs to buy. Look for two smart labels on the carton: "pasture-raised" and "organic." Makes sense—if the hens eat smart foods, they produce smarter eggs. Not convinced? Notice the deep orange color of the carotenoid-rich yolk from a hen fed "real food" versus the pale yellow yolk from a caged hen fed fake food. If you see "omega-3 enriched" on the label, be sure it adds "from *marine* sources" (since, as you learned on pages 32–33, omega-3s from a sea source are better for your brain than those from a land source). A good reference for the best eggs to buy is Cornucopia's Egg Scorecard.

WHY EGGS ARE SMART

What Your Brain Needs	What Eggs Give
• Smart protein	• Smart protein
• Choline	• Choline
• Carotenoids	• Lutein and zeaxanthin
• Smart fats	• DHA, the smartest fat (if feed is enriched with marine-sourced omega-3s)

3. AVOCADOS ARE AWESOME

Avocado is the smartest fruit because it contains healthy fats, protein, fiber, antioxidants, and other nutrients your brain likes. For the carb-cutters, avocados are virtually carb-free.

"But they're mostly fat," our fat-phobic friends claim. Yes, smart fats! Repeat, repeat, repeat: You don't get fat from eating smart fats, you get fat from eating dumb carbs (see why, page 74). And their high fat content gives them a rich, tasty mouthfeel that adds zest to smoothies, omelets, salads, or fajitas.

Here are some other reasons science says avocados are a supersmart choice:

More filling. Avocados have the highest protein content of any fruit. Because of this high protein content, plus their high fiber and healthy fat content, they tend to be more filling than other fruits. One day when you're hungry, try eating half of an avocado. The next day, at the same time with the same hunger, try eating a whole apple. While both are healthy, the increased fat, protein, and fiber in the avocado helps you feel fuller longer.

Cholesterol-friendly. People with high cholesterol levels showed a decrease in total cholesterol and LDL cholesterol, and an increase in HDL ("good") cholesterol, after boosting their avocado consumption.

More synergy. Avocados are the top *synergy* fruit, meaning avocados help extract nutrients from the foods they're partnered with. For example, adding an avocado to a salad increases the absorption of carotenoids like beta carotene, lycopene, and lutein in the salad's veggies (besides adding a delicious taste). This is why we advise adding avocado or avocado oil to smoothies, salads, and other dishes high in veggies. And that's not the only reason to eat avocado with your meal. Because avocados are full of fiber and very low carb, they blunt sugar spikes when paired with other, higher-carb fruits.

4. OH, THAT OLIVE OIL!

Extra virgin olive oil (EVOO) merits a high place on this list because it is the highest of any food in the top monounsaturated fat in the brain, oleic acid. Yet because it is low in the smartest fats, omega-3s, it is not as smart as seafood.

As you learned, the brain is most vulnerable to illness because of two main properties: the brain is mostly fat, which *oxidizes* (turns rancid), and the brain is a hypermetabolizer, so it generates lots of *oxidants*. Monounsaturated fats like oleic acid are less damaged by oxidation than other fats. Another perk of EVOO (and olives) is that it contains a powerful antioxidant/anti-inflammatory nutrient called *oleocanthal*.

Olive oil is also smart because it has been shown to lower the level of excess sticky cholesterol in the blood. EVOO is not only neuroprotective but also vessel protective: since the brain is one of the body's most vascular organs, what's smart for the blood vessels is also smart for the brain.

How much is smart? We suggest getting at least one tablespoon a day. When buying olive oil, look for a dark glass bottle (light can cause oxidation) and these smart words on the label: *extra virgin*, *organic*, and *cold pressed*.

SCIENCE SAYS: EVOO IS SMART

When surveying the medical studies, here's what we found: people who predominantly eat a Mediterranean diet—high in seafood, vegetables, and EVOO—suffer less dementia, meriting its distinction as a "neuroprotective" diet.

5. GO NUTS!

Being a "health nut" is good for your brain. What do nuts have in common with the previous supersmart foods? You guessed it—smart fats! Around 90 percent of the fat in nuts are the brain-friendly monounsaturated type.

One of the nutritious perks of snacking on nuts is *they don't create a sugar spike*. Another is that nuts have a *high satiety factor*, meaning they fill you up and satisfy your hunger with fewer calories and for a longer time.

Smart snacking on nuts is good for the brain. Rather than focusing on the "Top Nut Award" (the nut with the most nutrients per serving), which would be the awesome *almond*, we prefer you enjoy the nutritional perk of *synergy* (see page 62) by making your own trail mix, where each nut has some nutrients that are higher than others. Here is a brief course in "nutology" showing how each nut contains unique nutrients.

Nut to Add to Your Diet	Top Nutrients It Contains
Almonds	Fiber, calcium, vitamin E
Brazil nuts	Selenium
Macadamia nuts	Healthy fats
Hazelnuts	Tryptophan
Pecans	Fiber, thiamin, vitamin E
Pine nuts	Zinc
Pistachios	Lutein and zeaxanthin

Nut to Add to Your Diet	Top Nutrients It Contains
Peanuts	Protein, niacin, folate
Walnuts	Omega-3s

We suggest you shoot for a *palmful* (about 1 to 1.5 ounces) of raw or dry roasted (not oil roasted) nuts per day. Add a tablespoon of nut butter (peanut or almond) to your daily smoothie or grated cashews to your cereal, nibble on nuts that you carry in a baggie for mid-morning and mid-afternoon snacking, or top your evening salad with walnuts, pine nuts, or pistachios.

EAT SMART SEEDS

Seeds have a nutritional profile similar to nuts. They're rich in smart nutrients, such as the antioxidants vitamin E, selenium, and folate. Along with nuts, smartly sneak in to your daily cuisine seeds such as pumpkin, sunflower, flax, and sesame. Seeds are rich in the antioxidant *magnesium*, a nutrient in which many people are deficient. Dr. Bill grinds a tablespoon of pumpkin seeds for his morning smoothie.

A BIG, FAT, SMART CHANGE

Notice our top five supersmart foods all have one nutrient in common: they are high in (smart) fat.

Remember the two main goals of our supersmart foods are to *nourish* and *protect* your brain. While the previous five smart foods do both, mainly because they are full of smart fats and antioxidants, more nourishment and protection is needed from two more smart foods: berries and greens.

Because there are so many "players" in your brain's symphony orchestra (neurotransmitters, myelin makers, brain-cell growers, and so on),
and each one needs its own special neurochemicals for nourishment and
protection, the larger variety of neuroprotectants you get, the better. Or
as Dr. Mom taught: put more color on your plate. Each color provides its
own unique antioxidants, so the more colorful your meal, the more of
your brain's oxidants you "anti."

6. ENJOY A BERRY SMART BRAIN

Berry blends (blueberries, strawberries, raspberries, blackberries, and cherries*) are a combination of good taste, good science, and common sense. To
combat all the exhaust (oxidants) your hypermetabolizing brain puts out,
you need mouthfuls of antioxidants. Berries are loaded with them!

The "brain berry." Of all the berries, blueberries have the highest level of
antioxidants in their skin. The deep blue color, and the main neuro nutrient that keeps you from feeling blue, comes from the antioxidant *antho-
cyanin*. In animal studies, blueberries increased the concentration of one
of the "happy hormones," the neurotransmitter dopamine. Thanks to its
high antioxidant effect, in neuro-foodie circles blueberries have earned
the tag "the brain berry."

One day Dr. Bill was giving a supersmart foods talk to an audience
filled with children and labeled blueberries as "the brain berry." After his
talk, seven-year-old McKinney handed him her smart note as she said,
"Dr. Sears, I will always remember the brain berry."

Blueberries aren't just good for the brain; cardiologists also call them
the "heart berry." The nutrients in blueberries promote vasodilatation,
or blood vessels' ability to widen in size when needed to allow blood to
flow more efficiently everywhere, including the brain. Remember, one of
our top health tips is: the better the blood flow, the better the health of
every organ.

---

* Although cherries are not technically berries, we've included them here not only
because they are added to many berry-blend packages, but also because cherries, especially tart cherries, are rich in antioxidants.

More berry good berries. While blueberries are our top berry, others are a close second. Raspberries and blackberries are highest in fiber. Strawberries are highest in vitamin C. Cranberries are second only to blueberries in high antioxidant content.

Science says that berries:

- keep the BBB healthy.
- facilitate blood flow by keeping arterial linings smooth. Remember, the "plants" in your brain garden grow healthier the more blood flow they get.
- help speed up flow of traffic among nerve cells.
- protect the brain from the wear and tear of neuroinflammation.

Blueberries are also a favorite food for the bacteria living in your gut, which we'll learn more about in chapter three.

How to eat. Raspberries, strawberries, and blackberries must be bought *organic*, mainly because their skin, unlike the smooth skin of blueberries, is hard to wash. Each morning Dr. Bill puts a handful of a frozen organic berry blend (available at nearly all smart supermarkets) in his supersmart smoothie. He always pictures his brain and his taste buds registering "like!"

7. GREENS ARE GREAT FOR BRAIN HEALTH

About the only foods you will find that 100 percent of neuroscientists and food scientists agree on are veggies and greens. A summary of what science says:

- The more veggies you eat, the longer and the healthier you live.
- The more veggies you eat, the lower your risk of dementia and Alzheimer's. Two respected studies, the Chicago Health and Aging Project and the Nurses' Health Study, showed that people who ate more vegetables enjoyed a slower rate of cognitive decline.
- The more veggies you eat, the healthier your blood flow to vital organs like the brain.

- The more veggies you eat, the leaner your waist. And the leaner your waist, science adds, the larger and smarter your brain.

Why are greens so good for you? Again, it's Dr. Mom's wisdom to "put more color on your plate." The rich green color in spinach, kale, chard, collard greens, and asparagus is just what the brain doctor ordered, and the greener the greens, the more smart nutrients in them. Greens such as spinach and kale are rich in lutein and zeaxanthin, antioxidants that protect your oxidant-prone "fathead."

Besides being rich in antioxidants, there are two other main reasons why a salad a day helps keep the doctor away.

Veggies are "free" foods. Veggies are the only food category that can boast this recommendation: "Eat all you want." Veggies such as celery, artichokes, broccoli, asparagus, and greens require lots of chewing, which improves digestion, is good for your gut, and helps you feel fuller faster. In fact, if you chew in the way that's best for you (twenty to thirty times per mouthful), the muscular action of chewing plus intestinal digestion are likely to use up most of the calories that are in the veggies.

Salads don't spike. The carbs in salads are known as *slow carbs*, which is just what the brain prefers. Our brains perform better on a steady supply of carb fuel, rather than a spiky one, and the high fiber of the veggies blunts sugar spikes from carbs. Adding fatty foods to your salad, like a fillet of salmon, a tablespoon of olive oil, half an avocado, or some walnuts, further blunts carb spikes. (Fiber causes the carbs to be absorbed more slowly.)

How to get your greens. To make the smartest salad, enjoy the eating principle of *synergy*. Blend three to four different greens—such as Swiss chard, kale, spinach, and arugula—together in your salad, as each one has its own special nutrient profile. Even romaine lettuce, while not as good as these other four greens, is high in vitamin C and folate. A handful of kale or spinach is also great in a supersmart morning smoothie.

Enjoy what Dr. Bill calls "my big, fat Greek salad"—a spinach/arugula green salad, topped with a fillet of wild salmon, a tablespoon of drizzled olive oil, half a diced avocado, a tablespoon of hummus, a palmful of diced olives, and a few squares of feta cheese. Yum!

To boost nutrient absorption of your salad, give it some steam! Light steaming (around three minutes) increases the bioavailability of the nutrients in some vegetables, such as artichokes, asparagus, and beets. Several evenings a week Dr. Bill enjoys a "hot salad."

YOUR PERSONAL *FARMACY*

Leading functional-medicine specialist Bryce Wylde, author of *Power Plants*, once commented during a dinner next to the Sears' garden: "Bill, most of the *medicine* you need is right here in your backyard."

WHAT THE HAPPIEST PEOPLE ON EARTH EAT

Yes, there really are "happy meals" that are smart for your brain. Neuroscientists have concluded that there are three traditional diets associated with the happiest and healthiest people: the Nordic diet, the Japanese diet, and the Mediterranean diet. What these three diets have in common are four categories of foods that we prescribe in our healthy brain plan:

- safe seafood
- antioxidant-rich berries, especially blueberries
- antioxidant-rich vegetables, such as greens
- a variety of nuts

When we counsel patients with brain-health problems in our medical offices, we give them the memorable phrases: "Go fish!" "Go blue!" "Go green!" "Go nuts!"

A SECOND HELPING OF SUPERSMART FOODS

While we have listed the top seven foods that we believe, and science supports, are the smartest superfoods for your brain, there are many other foods that are also smart, and some are smarter than others. For example, nearly all veggies and other foods you can grow in a home garden are good for your brain. Here are some other smart foods that fuel good brain health, listed alphabetically:

Food	How They Contribute
Artichokes	Rich in vitamin C and folate.
Beans	Rich in protein and fiber. Navy beans are rich in choline.
Beets	Rich in natural nitrates that improve blood flow to the brain.
Broccoli	Rich in the powerful antioxidant sulforaphane. The whole family of crucifers—broccoli, kale, cauliflower, and so on—is especially good for your gut. Think "broccoli brain"!
Chocolate (especially 85% dark chocolate)	Rich in antioxidants, and improves insulin sensitivity and blood flow to the brain.
Coconut oil	Rich in MCT (medium-chain triglycerides), one of the favorite fats of both the gut and the brain (see Endnotes for why).
Coffee and tea	Rich in the powerful antioxidant phenols.
Garlic	Rich in antioxidants.
Ginger	Rich in antioxidants.
Lentils	Rich in protein and fiber, which are especially good for your gut.
Mushrooms (especially shiitake and crimini)	Rich in tryptophan, selenium, and antioxidants.
Pomegranates	Rich in antioxidants.

Food	How They Contribute
Red peppers	Rich in vitamin C.
Sardines and anchovies	Rich in DMAE, a nutrient that increases levels of the neurotransmitter acetylcholine.
Tomato paste	Rich in vitamin E and the antioxidant lycopene.
Turmeric	Rich in antioxidants, meriting our "smartest spice" award. Cultures that eat cuisines rich in turmeric have the lowest incidence of Alzheimer's.
Venison, wild game, and grass-fed beef	Rich in protein, iron, and omega-3s, and have much healthier fat profiles than feedlot-fed beef.
Yogurt and kefir (organic, 100% grass-fed, whole milk, and plain)	Rich in probiotics, vitamin K_2, and a smart fat, GLA—all smart ingredients for brain health.

You'll notice that many of these foods are similar to those on the list of smartest foods for your gut in the next chapter. As you will learn on page 88, the foods that are smart for the gut "brain" are usually also smart for the head brain.

Brain Health Farmacy

MORE SMART NUTRIENTS YOUR BRAIN NEEDS

Below are lesser-known but smart nutrients you need for brain health:

Nutrient	Smartest Food Sources	What It Supports
Iodine	Seafood, eggs, seaweed	Thyroid function, brain energy, myelination
Iron	Shellfish, pumpkin seeds, tofu, spirulina, grass-fed meat, wild game	Neurotransmitter function, attention, oxygenation, myelination
Copper	Oysters and other shellfish, Brazil nuts, beans and other legumes, mushrooms	Myelination, neurotransmitter health, antioxidant protection
Zinc	Shellfish, herring, eggs, nuts, legumes	Neurotransmitter function
Selenium	Shellfish, Brazil nuts, lentils	Antioxidant protection, thyroid health

SELECTING SMART SUPPLEMENTS

Before we get to how to eat all these supersmart foods, let's take a moment to talk about supplements.

If we all lived in a perfect nutritional world—one where we grew our own food on our own farm, and ate only what we grew, fished, or hunted—we probably wouldn't need nutrition in a pill. "Supplements" are just that. They fill in the nutritional gaps that exist in even the smartest diet.

The more you eat out of a box, the sooner you end up in a box.

Dr. Bill

We modern eaters suffer from what nutritionists call "micronutrient deficiencies." This means that because we deprive our plants and animals of good care and good feeding, modern produce and meat, like modern humans, are nutritionally deprived. Even worse, on its way from farm to fork, our food gets weakened with artificial processing and chemicals to increase its "shelf life."

In the previous chapters you learned that the two top nutritional needs for the brain—your "fathead"—are omega-3 fats (the smartest fats for your brain) and antioxidants (natural anti–wear and tear nutrients). To smartly select any nutritional supplement, remember *Dr. Bill's triad*:

1. Show me the *necessity*—do I need it?
2. Show me the *science*—what's the proof?
3. Show me the *source*—where does it come from?

"Do I eat enough seafood in my daily diet?" Hold up your fist. Do you eat a minimum of two fistfuls of wild salmon (or its equivalent in other seafood, like sardines or anchovies) a week? Probably not! If you don't, then you need an omega-3 nutritional supplement. (See "Endnotes" for the omega-3 contents of safe and popular seafood.)

"Do I eat enough fruits and vegetables every day?" Do you eat a minimum of ten fistfuls? Also unlikely. Therefore, you need to take supplements that are rich in the antioxidants found in fruits and vegetables.

Do you eat 10 servings a day?

Do you eat 2 servings a week?

SHOW ME THE SCIENCE

As a smart supplement taker, you want to know the science behind your supplements. First, has what's in the supplement capsule been proven to be bioavailable—that is, does it "get into the body," and from your gut into your blood? (We doctors joke about patients coming in with their bag of supplements that "go in the top end and come out the bottom end" without ever getting into the body.) And, second, have the healthful effects of the supplement been proven?

Good news: Omega-3s are the most scientifically researched nutritional supplement on the planet. More than 22,000 scientific articles have proven that the more omega-3s you eat *or take*, the healthier nearly every organ in your body will be.

The science is slimmer when it comes to most fruit and vegetable supplements. Nearly all fruit and vegetable supplements lack research into bioavailability. So, when selecting a fruit and vegetable supplement, look for scientific studies published in credible medical journals showing that hundreds of people were given the supplement and then had their blood drawn. The level of antioxidants found in their blood should go up to match the supplements they put in their mouths. The study should then follow these supplement-takers and discover the good health effects of these supplements, such as less inflammation and better blood flow.

SHOW ME THE SOURCE

Another thing to consider when it comes to selecting a supplement: you want it to come from clean soil and clean sea. The trusted sources we mention next are those whose fishing grounds and farms we have personally inspected to see how the seafood and plant food is fed and cared for.

CHECK WITH YOUR DOCTOR FIRST

While all the supplements we mention here have been proven safe and effective for most people, you may be taking certain medications or have an illness for which your doctor may want to change these recommendations. We strongly advise you to check with your healthcare provider before taking nutritional supplements, especially if you are currently taking prescription medications.

OUR FABULOUS FIVE SMART SUPPLEMENTS

1. **Omega-3.** Usually from a fish oil source, some are now available from sea plants for those looking for a *vegan* source.
2. **Fruit and vegetable concentrate.** A combined extract of many fruits and vegetables enjoys the nutritional perk of *synergy*. When you team many colorful foods together, as you do in a salad, smoothie, or in this case a capsule of dozens of fruits and vegetables, each nutrient enhances the effect of the others.
3. **Astaxanthin.** The natural nutrient that makes salmon pink, astaxanthin is one of nature's most powerful antioxidants and has been shown to help protect against dementia. The

science we really like is how
it has been proven to cross
the blood–retinal barrier
(that selective brain barrier
you learned about on page 16).
Neuroscientists assume that if it crosses
the eye barrier, it should be able to cross the
brain barrier as well, and get into brain tissue
where you want it to work.

I love astaxanthin!

Following our theme that good science and good sense are
trusted partners in health, follow this logic:

> *The brain is one of the top organs prone to oxidation (wear and
> tear that leads to disease).*
>
> *Astaxanthin is one of the top antioxidants that science
> shows, when eaten, gets into the brain.*
>
> *Therefore, to prevent wear and tear on the brain, eat or take
> more astaxanthin.*
>
> *That makes sense!*

Unless you eat two to three fistfuls of wild salmon each week
(you need a minimum of 6 ounces of wild salmon at least three
times a week to get enough astaxanthin to quell neuroinflammation), we recommend this supplement, especially for athletes
and vigorous exercisers.

4. **Vitamin D.** Besides our two favorite sources, salmon and
   sunshine, recent research shows that taking a vitamin D
   supplement can be an effective tool in preventing a wide
   range of illnesses including many cancers, diabetes, dementia,
   Alzheimer's, and others.

5. **Probiotics.** See chapter three to see how probiotics help make
   you gut brain smarter, and why your head brain also likes them.

SUPPLEMENT SOURCES WE RECOMMEND

See AskDrSears.com/supplements for the updated list of trusted sources we recommend and the dosage you need to take for the health needs you have.

I need it, but don't eat it, therefore I must take it.

HOW TO EAT SMART

Besides eating supersmart foods, the *way you eat* those foods also can affect your brain health. Here are the top five tips we prescribe in our medical practice.

FIVE FOOD RULES FOR HOW TO EAT SMART

How to Eat	Why It's Smart
Enjoy a brainy *breakfast*.	A breakfast loaded with the smart foods we recommend helps your brain feel calmer and more alert. It also programs the brain to eat smart the rest of the day.
Graze on mini-meals.	Grazers tend to be leaner and enjoy healthier brains and guts.
Eat *less volume*, but more nutrient-dense foods.	This lessens neuroinflammation, letting you live longer and healthier.
Balance your *omegas*.	Eating the right proportions of omega-3s and omega-6s balances your brain.

How to Eat	Why It's Smart
Eat the *real food* diet.	Lessens neuroinflammation, leaky brain, and leaky gut. Shapes your tastes toward liking real food and developing a distaste for fake food.

1. BEGIN YOUR DAY WITH A BRAINY BREAKFAST

Eating a brainy breakfast gives you a head start on a healthier day. The smart foods you eat at your first meal *frontload* your brain by sending the message, "Hey, brain, here's the smart way to eat the rest of the day." In contrast, people who begin the day with a high-carb, processed "out of a box" breakfast are more likely to "eat dumb" the rest of the day. Even dumber is skipping breakfast, which programs your brain toward *catch-up overeating*, especially by craving junk food and dumb carbs, the rest of the day.

> Once I cut out "crappy carbs" for breakfast, I
> felt so much better the rest of the day.
> –Erin Sears Basile, coauthor, *The Dr. Sears T5 Wellness Plan*

Make breakfast a "happy meal." Beginning the day in a happy mood from eating smart foods programs your brain to eat smarter the rest of the day. On the other hand, high levels of the stress hormone cortisol—released when you gorge on junk carbs—trigger the production of a neuro-hormone called NPY, which triggers carb cravings and sets you up for unhealthy eating. Remember, happy meals produce happy hormones; stressful meals produce stress hormones.

Give brains a smart start. A brainy breakfast gives you a head start—literally. Here's a summary of what science has found for young brains, with implications for the rest of us to smartly start our workday. Smart young breakfast eaters:

- are more attentive in class.
- make higher grades and get better reading and math scores.
- are less likely to be diagnosed with learning disabilities.
- show improved memory.
- experience less depression, anxiety, and hyperactivity.

The Smartest Ingredients Of A Brainy Breakfast

Powerful proteins. Proteins help jumpstart a brainy day because they stimulate the release of the "alertness" neurohormones norepinephrine and dopamine. Proteins are also rich in the amino acid tyrosine, a neurostimulant that is a building block of norepinephrine and dopamine. Another smart nutrient in proteins is the amino acid tryptophan, a building block of serotonin, the calming neurohormone. Remember the two T's: tyrosine and tryptophan. They *balance the brain* by revving it up to be alert, but also calming it to avoid anxiety and brain drain.

Smart fats. Pairing protein with healthy fat helps comfortably fill you up—both have what's called a *high satiety factor*—and keep you fuller longer. This is why eggs (rich in both healthy fats and protein) are our vote for the second smartest breakfast food. Our supersmart smoothie (see page 73) is our first pick. Seafood is another good choice. As you learned, seafood is the smartest food, and we wish Americans could enjoy the smart habit of eating seafood for breakfast like many Asian cultures do.

Fiber-filled, slow-release smart carbs. Smart carbs are your brain's favorite fuel. Think of smart carbs as time-release energy packets that give the brain a *steady* supply of fuel. Brainy breakfast carbs are partnered with protein, fiber, and fat, three nutrients that "hold hands" with carbs when they reach the intestines to keep the carbs from being absorbed and used too fast. Too-fast absorption and use leads to a low blood sugar crash—the cause of the typical mid-morning brain drain after eating a high-carb, and especially a dumb-carb, breakfast.

"INSTANT" ISN'T SMART

Avoid most "instant" foods for breakfast. The carbs in instant foods go "instantly" into the brain and are more likely to be used up too quickly. In fact, a study at Children's Hospital Medical Center of Boston, where Dr. Bill did some of his medical training, compared the blood sugar of children eating instant oatmeal with those eating steel-cut, regular oatmeal. Those who ate the unprocessed, non-instant oatmeal showed a slower rise and fall in their blood sugar levels and felt more satisfied throughout the morning. In contrast, children who ate instant oatmeal had more rapid changes in their blood sugar and tended to eat higher-calorie snacks the rest of the day.

Brainy Breakfast Meals

A few of our favorite brainy breakfast meals include:

- a veggie whole-egg omelet with guacamole
- crockpot oatmeal with berries
- organic, 100 percent grass-fed, whole-milk, plain yogurt with berry medley
- a smoothie (see page 73)

See our recipes for more brainy breakfasts at AskDrSears.com/recipes.

2. GRAZE (AND SIP) TO YOUR BRAIN'S CONTENT

Grazing is good for the brain because of three S's: *slow* and *steady* release of nutrients that nourish the brain, and no *spikes*. Simply put, your brain needs your bloodstream to deliver a *steady* supply of the right fuel—not so much that you get blood sugar and blood fat spikes, but not so little that the brain runs out of fuel or gets hypoglycemic.

Grazers think clearer. When your body is trying to quickly clear sugar spikes, it sometimes sends your sugar level too low, which dulls thinking.

Grazers eat smarter. Going too long between meals can prompt you to feel hungrier and therefore make dumb food choices, such as junk carbs. Grazing when you are "pre-hungry" (feeling just *slightly* hungry) instead prompts you to make smarter food choices.

Grazers tend to be leaner. Because grazers eat smarter, they also tend to be leaner. Throughout our book you will learn the relationship between leanness and smartness. (See what science says about the relationship between a leaner belly and a smarter, longer life, page 102.)

SCIENCE SAYS: GRAZING IS GOOD FOR THE BRAIN

Here are just some of the top health effects that research on grazing has shown:

- lower risk of Alzheimer's
- steadier, happier moods
- improved cerebral blood vessel health
- less inflammation
- fewer "highs": high blood sugar, high blood cholesterol, and high blood pressure—all of which have been shown to be harmful to your brain.

TAKE TIME TO SEE AND SMELL THE MEAL

The pleasure of eating begins in your brain. Remember coming home and smelling Mom's cooking? Even before taking your first bite, feast your eyes and nose on the food. Your brain can start registering "like" even before your tongue does, as it reflects on the smart nutrients contained in the food you see and on all their brain-health benefits. This exercise helps train your brain to crave the foods that are good for it, even though they are foods that you may have previously disliked.

Great Grazing Tips

Every day in Dr. Bill's medical practice he writes on a prescription pad what he calls the "rule of twos":

- Eat *twice* as often.
- Eat *half* as much.
- Chew *twice* as long.

Why chew twice as long? In gut speak, the more work you do in the top end, the less work for the bottom end. Finely chewed food is easier to digest (see more about how chewing is good for the gut, page 90). Plus, chewers enjoy their meals more so than do gorgers. Let the food linger in your mouth while enjoying its "mouthfeel." Keep chewing until the food is mostly liquid and slides down easily without much conscious swallowing. Many fibrous foods, like broccoli, require forty chews. Savoring a bite of salmon may only require thirty.

Besides grazing, the gut loves sipping too.

The Sipping Solution

We believe that a daily smoothie full of brain foods is one of the smartest brain-healthy changes you can make. Dr. Bill began drinking a daily smoothie in 1997 and has continued this smart sipping solution ever since.

Two reasons why a smoothie a day can help keep the brain doctor away are:

1. **It's berry good for your brain.** Remember antioxidants, the class of neuroprotective foods you learned about on page 13? Our supersmart smoothie is loaded with foods like blueberries—the "brain berry"—that are full of these anti–wear and tear, anti-inflammatory, anti-rust nutrients.

2. **It balances blood sugar in your brain.** One reason we call it a "smoothie" is it can *smooth blood sugar spikes*. Our smoothie is full of foods naturally low in fast-release sugars and high in fiber and healthy fat, which slow the release of those sugars. *Low* and *slow* is just what the brain doctor ordered. We add another word to low and slow: *steady*. Unlike other high-fuel-using tissues like muscles, the brain does not store sugar, so it needs a steady supply throughout the day. Sipping is simply fluid grazing.

A BIG FAT SHAKE MISTAKE

The biggest mistake shake makers make is they unwisely leave out healthy fats. Fats increase the absorption of fat-soluble nutrients, such as vitamins A, D, E, and K, as well as carotenoids and flavonoids. Also, fats give your drink a savory mouthfeel.

---

3. EAT LESS, LIVE SMARTER AND LONGER

Neuroscientists are proving that eating less translates to living lon-ger and having a longer brainspan. It's an eating change called "caloric restriction," but we prefer to describe it as "eating smarter calories." Cut-ting down on calories doesn't—and shouldn't—mean cutting down on nutrients. Proper caloric restriction, especially that which is scientifically proven to have smart brain effects, only cuts down on calories that the brain and body don't need.

Why eating less translates to living healthier is still being studied. Our take is the brain and body become *more fuel efficient*, like the way your car engine lasts longer and runs more efficiently when you use a cleaner-burning fuel. The more efficient the fuel, the less exhaust is pro-duced, so your brain tissue sustains less wear and tear.

Eat less, stress less. Eating less tends to down-regulate glucocorticoid receptors, which means these receptors become, shall we say, less recep-tive to the potentially damaging effects of excess circulating stress hor-mones. We tend to eat more empty calories when we're stressed—just the opposite of what the brain needs us to do.

Eat less, make insulin behave better. Eat-ing less (without sacrificing feeling satisfied) also increases *insulin sensitivity*, which means insulin uses glucose more efficiently—like getting more mileage per gallon from your fuel. Increased insulin sensitivity (efficiency) ensures a steady supply of glucose fuel to the brain, but with lower insulin levels. Insulin *in*sensitivity, or type II diabetes, is the fastest-growing illness in America, and possibly worldwide. It's also, as you will learn in chapter eight, a leading contributor to Alzheimer's disease.

The smaller your meal the better I feel.

Persistently high insulin levels are one contributor to brain aging. As we get older, our metabolic rate and our insulin sensitivity lessens—making eating less, because of its effect on insulin sensitivity, one of the simplest solutions for healthier aging. *Lower insulin levels* have been shown to be associated with fewer neurodegenerative diseases.

Dr. Bill's Supersmart Smoothie Prescription

- Choose from each of the seven food categories shown in the illustration on the opposite page. Begin with the ingredients that you already know you like and gradually try the others. Eventually your *gut feel* will decide the best recipe for you.

- Be sure to add *protein* and *healthy fats* to each smoothie, which will taste better and keep you fuller longer than a carb-only shake. Our basic recipes (found at AskDrSears.com/Smoothies) have calorie percentages of 20–25 percent protein, 30–35 percent healthy fat, and 40–50 percent carbs.

- Begin with one 12-ounce smothie at breakfast two or three days a week, then gradually increase to a smoothie a day, five days a week.

- Gradually increase the volume of the smoothie you make from 12 ounces (one glass) to as much as 64 ounces (five glasses, or the volume of most blenders). Sip on your 64-ounce smoothie in frequent mini-meals throughout the day, as breakfast, lunch, and mid-morning and mid-afternoon snacks. Then eat a healthy dinner.

Dr. Bill's Supersmart Smoothie

Healthy Liquids
Coconut milk, Almond milk, Goat milk*,
Kefir**, Green tea, Organic juices: green,
vegetable, pomegranate

Healthy Proteins
Organic Greek yogurt**, Nut butters,
Nuts, Hemp powder

Healthy Protein Powder,
Multinutrient Mix
Juice Plus Complete, chocolate or vanilla

Healthy Fats
Avocado, Coconut oil: virgin, Coconut chunks,
Nut Butter, Ground flaxseed or chia seed

Healthy Carbs
Berries***, Strawberries, Pomegranates,
Papaya, Kiwi, Banana

Greens
Kale, Spinach, Chard

Special additions
Cinnamon, Grated ginger, Wheat germ,
Figs/ raisins/ dates, Cacao, Mint,
Hawaiian spirulina

*Goat's milk is higher than cow's milk in MCTs (one of the brain's favorite fuels; see page 57 for more MCT info).

**Organic, 100 percent grass-fed, plain, whole milk

***Our favorite is an organic blend: blueberries, strawberries, raspberries, and cherries.

Eating Smarter To Eat Less

Eating smarter naturally translates to eating less. Why? Mainly because of the smart fats and high fiber, two fill-up-faster ingredients in our healthy brain plan.

Try this test: For a couple weeks, just eat the smart foods on our list (page 30). You will be eating from the periphery of your supermarket, not the boxes and bags in the center aisles. You are likely to feel fuller faster, while eating *less* volume. Both your brain and gut will thank you.

Don't eat dumb. Dumb foods are the reverse of smart foods. They are low in fiber, high in spiking sugars and fats, and contribute nothing but unhealthy effects on your body and brain!

The dumbest foods are chemical food additives, artificial sweeteners, taste enhancers, and artificial colorings and flavorings. Noted neurosurgeon Dr. Russell Blaylock, in his book *Excitotoxins: A Taste that Kills*, exposes the possible brain-damaging effects of many of these chemical food additives, which neuroscientists have dubbed *excitotoxins*. When neurons are exposed to these artificial chemicals, they become so excited and fire so rapidly that they quickly become exhausted. Some even die. (See "Endnotes" for an in-depth exposé of artificial sweeteners and flavor enhancers.)

An example of eating dumber to die sooner is the usual "chips and sips" and "fries" that infect teen and young adult eating. Studies show that people who eat MSG-filled fries are five times more likely to continue eating this "duo to die for," because they changed their brain to crave these junk carbs.

Don't do diet drinks. Some forms of excitotoxins are more dangerous than others—like the ones in diet drinks. Liquid excitotoxins are much more toxic to the brain than dry forms because liquid ones are absorbed faster and produce higher blood levels.

If you needed even more reasons to avoid diet drinks, brain doctors are now teaching that artificial sweeteners are actually more harmful for the brain than sugar. Here's why. Remember that the hormonal symphony in your brain is designed to protect it. When we eat a piece of

Mom's homemade apple pie, the hormones in the brain play sweet music, saying: "I like this." But eventually, another set of hormones dial down the sweet craving and you put your fork down. Artificial and refined carbs don't enjoy this dial-down mechanism. Diet drinks provoke the "eat more" hormone but don't ever turn off carb cravings. There is a reason that you no longer see commercials advertising use of diet drinks for weight loss. Several studies showed that switching to diet drinks caused most research subjects to actually gain weight over a three-month period.

THE NEUROTOXIN NAUGHTY LIST

- Artificial sweeteners
- Chemical food colorings
- Chemical flavor enhancers
- Hydrogenated oils
- Cottonseed oil
- "Natural flavors"
- "Instant" carbs
- MSG

Eat fat and fiber first—not carbs. The worldwide eating custom of "serve salads first" makes sense. The longer chewing required of fiber-filled salads satisfies your hunger sooner, prompting you to eat less. And when you add fat, such as olive oil, olives, feta cheese, or a fillet of salmon, a salad can replace that big carb-filled pasta entrée.

In contrast, when you start with eating refined carbs, especially when the "carb effect" is enhanced by MSG, you stimulate the brain into carb-craving. These sugar cravings then prompt the hypothalamus, which controls your metabolism, to turn on the "eat more" hormones cortisol, adrenaline, and noradrenaline. There is a reason why restaurants give you free bread first! Not only do excess levels of these hormones lead you to eat more and dumber foods, they're also a major contributor to anxiety, depression, agitation, and aggression.

ENJOY HOME-MEAL ADVANTAGE

To preserve your brain, eat in more than you eat out. Excluding breakfast, the typical American eats more than 50 percent of their meals outside the home, and at restaurants we usually eat dumb foods like white bread, fries, and soda.

Move first, eat smarter. Doing at least one minute of exercise prior to ordering or eating stimulates *leptin*, which is the hormone that helps you feel full sooner and avoid overeating. (Think: leptin, eat *less*.) Even a brief walk to the bathroom or outdoors twenty minutes before your meal gives you a pause that allows your satiety hormones to prompt you not to overeat. Or try isometric exercises (see anytime-anywhere exercises, page 111).

"COMFORT FOODS"–NOT SO COMFORTABLE!

The fact that sweet foods are called comfort foods has a neurochemical basis. They trigger the release of the feel-good neurohormone serotonin. Yet unlike feel-good thoughts and experiences, which can last for hours, the neurochemical effect of a comfort food usually wears off within twenty minutes . . . prompting you to crave more comfort foods even though you may not still be hungry.

4. BALANCE YOUR OMEGAS, BALANCE YOUR BRAIN

We've talked a lot about omega-3 fats, but there's another omega fat you need to know about: omega-6, found in seafood and many oils. Both omega-3 and omega-6 fats are called *essential fats* because your brain can't live without them. Yet too much of a good nutrient, like omega-6s, can be bad for brain health. So it's important to keep these two smart fats in balance.

The "bully effect." Tissues, especially brain cell membranes, need both omega-6 and omega-3 fats. At healthy ratios—around a 1:1 or 2:1 ratio of 6s to 3s—we build healthy brain tissue. But when we eat too many 6s and not enough 3s, as in the SAD diet—which can be out of balance by as much as 10:1 or even 20:1—the excess 6s use up all the enzymes for their own metabolism, so there are not enough enzymes left over for the 3s. The result is an omega-6/omega-3 imbalance that can lead to neuroinflammation.

We call this the "bully effect." Think of omega-3s and omega-6s as friends who like to play together. Things go well as long as they play nicely. But excess omega-6s act like a bully, overpowering the omega-3s. The body is left with an omega-6 excess and an omega-3 deficiency that prompts your otherwise smart immune system to pump out excessive biochemicals, called *inflammatory markers* or, simply put, more sticky stuff that collects and infects your blood vessels and your brain. This *neuroinflammation* is the root cause of most brain diseases.

How to balance your omegas. The best way to enjoy and feel omega balance is to just eat real foods. When good oils went bad—when they were changed by food factories—we began getting too many omega-6 fats and not enough omega-3s. (Remember, it's the *excess* of the omega-6s that can cause neuroinflammation.) As you will learn in chapter seven, most mood disorders are due to a biochemical imbalance in the brain. Balance your omegas to balance your brain.

> **To put your brain at ease, balance your omega-6s and omega-3s.**

One of the most obvious ways to balance your omegas, and therefore your brain, is by balancing your oils:

GIVE YOURSELF A SMART OIL CHANGE

Eat Smart (Best blends of healthy fats)	Eat Less (Too high in omega-6s and too low in omega-3s)	Eat None
• Fish oil (supplements) • Algae oil (supplements) • Olive oil* • Coconut oil, virgin** • Flax oil • Hemp oil • Avocado oil	• Corn oil • Soy oil • Sunflower oil • Safflower oil	• Hydrogenated oil • Cottonseed oil*** • Canola oil****

*One tablespoon a day is our recommended serving.

**Coconut oil is naturally rich in MCTs (medium-chain triglycerides), which may improve cognitive function in patients with dementia and are healthy fats for intestinal health. Coconut oil was long unscientifically maligned as a "saturated fat," but it doesn't increase sticky stuff in the blood vessels the way saturated fats in meats do. New nutritional insights reveal that all saturated fats may not be unhealthful after all.

***Contains one of the highest pro-inflammatory omega-6/omega-3 ratios—greater than 200:1. Also likely to be contaminated with pesticides.

****Highly processed and chemicalized.

5. EAT REAL FOODS

Along with following these four tips, the best thing you can do for your brain health is eat a *real food* diet, which simply means eating nutrient-dense foods with:

When you eat real food, I feel real good.

- no processing that changes the nutrient genetics or molecular structure of the food.
- no chemicals added.
- no nutrients removed.

DR. BILL'S DAILY DIET

From the smart foods you have learned about in this chapter, you can see why most of Dr. Bill's days are spent eating this way:

- *Hydrating* his brain by drinking three 8-ounce glasses of water upon awakening (see why, page 184).
- *Frontloading* his brain shortly after awakening by sipping on a supersmart smoothie (see page 65).
- Enjoying a cup of coffee (half-caf) with a teaspoon of coconut oil (see brain-health benefits of coconut/MCT oils, pages 78).
- A couple days a week, enjoying a breakfast of a veggie omelet, spiked with turmeric and black pepper, served with guacamole. (Black pepper increases the antioxidant absorption from turmeric tenfold.)
- Snacking during the day by nibbling on nuts.
- Topping off his brain needs in the evening with a smart salad, paired three to four days a week with 4 to 6 ounces of safe seafood, usually wild salmon.
- Filling in the nutritional gaps by taking smart science-based supplements.
- Savoring an occasional treat: venison steak *Diane*, paired with a glass of zinfandel.
- Finally, some nights, pleasing his brain before retiring with one egg as a before-bed snack.

FOOD RULES FOR BRAIN HEALTH

- Just eat real foods, mostly plants.
- Savor safe seafood.
- Eat more vegetables.
- Take smart science-based supplements.
- Eat less "added sugar" and fewer refined carbs.
- Graze, don't gorge.
- Chew longer to live longer.
- Fix your gut to fix your brain—which you will learn about in the next chapter.

CHAPTER 3

BE GOOD TO YOUR GUT– YOUR SECOND BRAIN

F ix your gut to fix your brain. Surprised? Often the quickest way to change your brain is to change your gut. Your gut is wisely called "the second brain" because your head and gut are partners in mental health. With this partnership in mind, let's take a journey into your gut to learn how better gut health leads to better brain health.

YOUR TWO BRAINS ARE CONNECTED

Your head brain and gut brain are connected by the largest nerve highway in your body, the vagus nerve. Your head brain and gut brain send millions of biochemical text messages to each other through this nerve highway every day, monitoring each other's smart, or sometimes dumb, activities. This internerve connection accounts for the sayings "gut feelings," "gut instinct," "and deep inside I feel . . ."

If your head brain is happy, so is your gut brain. If your head brain hurts, your gut brain is likely to feel queasy.

The vagus nerve also has many branches, or side roads, off its main highway that go to many other organs, such as the heart, lungs, kidneys, and so on. This branching explains why when your head brain hurts, many of your other organs misbehave.

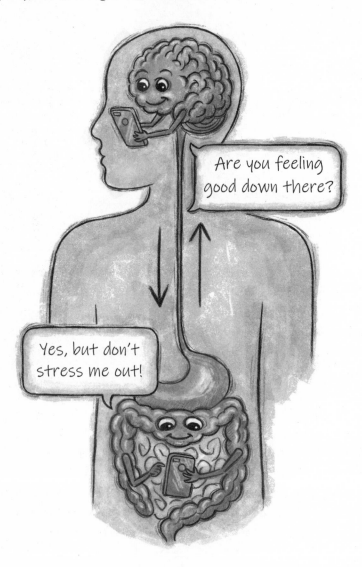

GUT BRAINS, LIKE HEAD BRAINS, LEAK

On page 14 you learned about how the blood–brain barrier (BBB) protects sensitive brain cells from attack by neurotoxins. When the BBB weakens, it leaks. Neurotoxins seep through and damage brain tissue by driving *neuroinflammation,* the root cause of most mental illnesses.

"Leaky gut" is another epidemic triggered by not-so-smart living and eating. The lining of your intestines also functions as a gut–brain barrier (GBB), and toxins can damage the GBB just like they do the BBB. Toxic foods and artificial chemicals in the modern SAD (standard American diet) can poke holes in the GBB, leading to "leaky gut"—the root cause of most intestinal inflammation, such as irritable bowel syndrome. In some ways a leaky GBB may be worse than a leaky BBB, in that harmful artificial chemicals that leak through the GBB can get into the bloodstream and inflame many other tissues of the body. Leaky gut is often the root cause of many of the *-itis* illnesses, such as colitis, thyroiditis, and arthritis.

> **Fix your gut lining, fix your brain *and* body.**

A mixed-up immune system. The total body effect of leaky gut is mostly due to a mixed-up immune system. When artificial molecules from artificial food chemicals leak through, the immune system goes on high alert: "What are these foreign invaders? Just in case they are bad for the body, we'll attack them." By a quirk of biochemistry, the antibodies told to attack the foreign chemicals that leak through get confused and target other tissues of the body, such as thyroid and joint tissue, as well. This immune system confusion, called autoimmune disease (a category that includes things such as thyroiditis and arthritis), is occurring at younger ages.

Allergies are thought to be the result of your overly sensitive immune system having an anxiety attack. Another example of immune system confusion is the growing incidence of "gluten sensitivity." When we no longer eat real wheat, we don't get a real smart-gut response. Because we have messed up the biochemistry of wheat through hybridization, a double-fault occurs: the messed-up wheat gluten causes gaps in the GBB, and then the foreign wheat molecules seep through. The immune system attacks not only the wheat molecules, but also other tissues of the body. This is also why many people with auto-immune illnesses must go on a *real food* diet and eliminate the messed-up molecules in foods such as wheat, dairy, and soy. (See related sections: "Gluten Brain," page 326, and; "Is Dairy Smart?," page 327.)

The SAD—standard American *disease*—of "I feel sick and tired" usually begins in the gut. You feel sick because your internal immune system army is not defending you in battle against germs and toxins; they're too busy focusing on foreign molecules in your food. You also feel tired because your internal energy batteries—your mighty mitochondria (see page 6)—aren't get-ting enough fuel to produce the energy you need.

AUTISM: ANOTHER GUT-BRAIN CONNECTION?

Pediatricians have long suspected that one contributor to the growing epidemic of autism could be an autoimmune disorder from a messed-up microbiome. Many of these children also suf-fer "pains in the gut," most often from eating "unreal" wheat, dairy, and soy. This is why Dr. Bill often begins each autism con-sultation with his "eat real foods" talk.

MEET YOUR MICROBIOME

Suppose you consult a top doc in preventive medicine for brain health. Let's call her Dr. Good Gut.

"How's your microbiome?" she asks.

"My what?" you may wonder.

Dr. Good Gut says, "The best way to have a healthy brain is to have a healthy microbiome. Most of the patients I see at all ages have a messed-up microbiome."

Microbiome, meaning "tiny bug home," is the name for the community of bacteria that reside in the lining of your large intestine. In return for free food and a warm place to live, they do smart things for your body and brain. Think of your gut bugs like the managers of your gut-brain health.

Your microbiome is the biggest zoo and the biggest social network in the world. An estimated 100 trillion bacteria reside in your gut, mostly in the colon, where more than a billion of these bacteria can live in just one drop of intestinal fluid. If you were to bunch all these bugs together in a jar they would weigh around three pounds, making the microbiome one of the largest "organs" in your body. The better you care for your gut bugs, the better they care for your brain.

Open your gut-brain pharmacy. Your microbiome is your *internal gut pharmacy*. These good gut bugs take the leftover food that your upper intestines couldn't digest, especially fiber, and make medicines out of it. Here's a list of what these good gut bugs do for your body:

- Make nutrients your body needs.
- Help keep your body lean.
- Balance the immune system.
- Defend against harmful bacteria that enter the gut.
- Grow a non-leaky gut lining.
- Help prevent allergies.
- Help prevent intestinal *-itis* illnesses.
- Ease pains in the gut.
- Lower risk of colorectal cancer.
- Mellow your moods.

Yes, you may be amazed to know that most of your "happy hormone," serotonin, is made in your gut brain—approximately 90 percent is made there, while only 10 percent is made in your head brain.

We make happy medicines

WELCOME TO THE WORLD OF "BIOTICS"

Three smart, but confusing, terms you will now see in most health and nutrition writings are *prebiotics*, *probiotics*, and the newest, *psychobiotics*. Here's what they mean and why you need them:

- *Prebiotics:* The *fiber* in food that ferments into the nutrients that feed your microbiome. Unfortunately, modern food processing removes much of the prebiotics.
- *Probiotics:* The billions of good gut bacteria available as a supplement, which helps populate your gut garden with healthy bacteria; the word means "pro-life."
- *Psychobiotics:* The gut bugs in your microbiome that produce neurochemicals (such as serotonin, dopamine, and GABA) that benefit the head brain and gut brain; a term coined by Ted Dinan, coauthor of *The Psychobiotic Revolution.*

What motivated me to continue to breastfeed my baby longer was when Dr. Bill—my dad—explained to me how I'm fertilizing my baby's gut garden for life, and to think of my milk as like "homemade kefir."
–Erin Sears Basile, new mother and coauthor of *The Dr. Sears T5 Wellness Plan*

YOUR MIND–MICROBE CONNECTION

We know that the head brain and the gut brain have an interpersonal relationship, where they keep each other accountable for their actions. But exactly how the gut's neurochemical text messages reach the brain is still uncertain. It's probably through several channels: the bloodstream, the vagus nerve, and the immune system.

What's clear is that your microbiome influences your mind. In some ways, the smarter the bacteria in your gut brain is, the smarter the tissue in the head brain becomes. Exciting new discoveries have revealed that the bacteria in your gut help your brain garden get the right amount of fertilizer by producing neurohormones that alter behavior and emotion and releasing them into the bloodstream to go to the brain.

This interconnection also means that if the microbiome is in a bad mood down there, the head brain can be in a bad mood up here. There could be something "bugging" you in your gut. So instead of your anxiety being "all in your head," it could also be "in your gut." Got it? Hence, the advent of the most exciting, yet mysterious, new science called *psychoneurogastroenterology*. Translation: Fix your gut to fix your brain.

A top theme of our healthy brain plan is *how to think-change your brain* from sad to happy. The growing science of the mind–microbe connection suggests you can also think-change the behavior of your gut bugs.

IS YOUR GUT BRAIN ALSO DEPRESSED?

"My microbiome made me feel it" may prove to be true. While research into this whole mind–microbe connection is still in its infancy, studies are revealing that your microbes may contribute to mood disorders. Known as the "Transferring the Blues" study, when researchers inserted feces (yes, poop!) from depressed people into rats, the rats also became depressed.

THE SEVEN SMART GUT CHANGES FOR YOUR BRAIN HEALTH

While writing this book, Dr. Bill had the gut experience of working on a microbiome project with expert microbiologists at the University of California, Davis, in addition to surveying what science says about the gut–brain connection. With this information, here are the seven changes he has successfully prescribed in his medical practice:

1. EAT GUT BRAIN-FRIENDLY FOODS

Our brain-health diet is also a gut-health diet! It includes these smart gut-friendly features:

1. *High* in easily digested soluble and insoluble fiber, or what we call "fiber-friendly foods." Think: "Fiber feeds my gut-health pharmacy."
2. *Low* in processed foods where the fiber and other rich nutrients have been processed out and artificial ingredients put in.
3. *High* in a rainbow of colors—red, green, yellow, purple— from fruits and veggies. Your microbiome feeds on colorful antioxidant-rich foods.
4. Organic/pesticide-free, when possible.
5. Blended or well-chewed for easier digestion.

Also, avoid foods with "chemical emulsifiers" (artificial chemicals that help all the ingredients mix better), such as polysorbate 80, which may thin the intestinal mucus and harm the microbiota.

Before you take a bite, think: "Is this food good for my microbiome?"

BEST FOODS FOR A BETTER BOWEL BRAIN

We have chosen these foods not only for their high fiber (both soluble and insoluble), but because they are rich in nutrients that feed a healthy microbiome. Here's our healthy microbiome traffic-light eating list:

Green-light foods (eat more)		
Apples (organic*)	Cottage cheese**	Pomegranates
Artichokes	Endives	Pears
Asparagus	Garlic	Pepper, black
Bananas	Ginger	Quinoa
Barley	Greens	Sauerkraut
Beans	Green tea	Squash
Bee pollen	Kale	Sweet potatoes
Beets	Kefir	Tempeh***
Berries	Leeks	Tofu***
Bone broth	Lentils	Vinegar: apple
Broccoli	Miso***	cider
Brussels sprouts	Nuts and seeds	Wild rice
Cabbage	Oats	Yogurt (see page
Cauliflower	Oils: olive,	58)
Cinnamon	coconut, avocado	
Coconut,	Omega-3-rich	
unsweetened	seafood	
(chunks and oil)	Onions	
Yellow-light foods (eat less, or "white out" your diet)		
White rice	Pasta with refined	Added sugar
White bread	grains	

Red-light foods (eat none; resist the "sinful seven")****		
Artificial sweeteners Artificial flavor enhancers (MSG and its aliases, page 75)	High-fructose corn syrup Sugar-sweetened, carbonated beverages Fast foods: fries, nuggets, burgers, hotdogs	Pesticide-sprayed foods (see the "dirty dozen" at EWG.org) Nitrite-preserved deli meats

*Go organic with all these foods when possible!
**Choose dairy products that have the smart four: organic, 100% grass-fed, whole milk, and plain.
***Insist on soy that is organic and non-GMO.
****Also see the neurotoxin naughty list, page 75.

2. CHEW-CHEW TIMES TWO

Remember Dr. Bill's "rule of twos" from the previous chapter:

- Eat *twice* as often.
- Eat *half* as much.
- Chew *twice* as long.

Your microbiome likes that! Besides releasing more saliva, which is rich in digestive enzymes, chewing slows down your eating so that more digestion occurs in the mouth and fewer big heaps of food reach your colon, making your gut's job easier.

Salivate more, leak less. A growing dental science that studies the *periodontal/systemic connection* is revealing fascinating facts about how what goes on in our mouths can influence, for better or worse, what goes on in our guts. Saliva is one of the first "medicines" your upper digestive tract produces. Besides sending a mouthful of friendly microbes into your gut alongside your food, saliva naturally lubricates your gut's lining. Our "chew-chew times two" advice increases the amount of saliva you make. My savvy dentist friends advise, "The healthier your mouth, the healthier your microbiome."

Microbiome memories. As Dr. Bill advises daily in his medical practice:

> *The better you chew,*
> *the better you poo.*
> *And, the brain feels,*
> *that's good for you!*

3. ENJOY THE SIPPING SOLUTION

The sipping solution is as good for gut health as it is for brain health (see page 70). For more than twenty years Dr. Bill has taught that better health begins with this one simple change, and he's found it to yield the quickest gut-health effects. It's especially helpful for easing two common "shuns"—indiges*tion* and constipa*tion*. Because the blender does most of the work at the top end, it's easier for the gut to do its work at the bottom end. And again, when your gut brain feels good, your head brain feels better.

BLEND, DON'T JUICE

Juicing removes most of your microbiome's favorite food, fiber, from fruits and veggies, while blending preserves it. If your gut brain and head brain could talk, they would say, "Blend, don't juice!"

---

4. DRINK LESS ALCOHOL, LEAK LESS IN YOUR GUT

Drinking too much alcohol too fast contributes not only to pains in the gut but a leaky gut. Drinking smart can help keep you feeling smart in your gut. Neurologists label alcohol as a "depressant" for the head brain, and new research suggests it works similarly in the gut brain. An experiment known as the "Drunk Bugs" study, published in 2017, showed that mice under the influence of alcohol suffered *depression* of their good gut bugs.

Besides getting your smart gut bugs drunk, excess alcohol damages the protective mucous lining of your intestines. Read chapter ten, "Drink Smart," to learn how to keep alcohol from damaging both your head brain and gut brain.

5. GO DRUGLESS!

Many prescription and over-the-counter medications, especially if used too strong for too long, can damage the intestinal lining and contribute to leaky gut. When possible, *get the ants out*: *ant*acids, *anti*-inflammatories, *anti*-depressants.

6. STRESS LESS, FEEL GOOD IN YOUR GUT

Ever notice when your brain feels upset, so does your gut? The medical specialty of *psychoneuroimmunology* looks at how meditation helps the mind prompt the bowel to feel better.

Why stress causes a pain in the gut. Stress slows digestion. It starts at the top: saliva secretion decreases, giving you a dry mouth and depriving the food you eat of the enzymes that facilitate digestion. Next, the stressed stomach slows down its mashing of that food and secretes less of its own digestive enzymes and acids. Finally, the rest of your intestines slows down its peristalsis, or snake dance, resulting in less food nutrients being absorbed. Undigested food backs up in the intestines, making you feel constipated.

Why does digestion slow down when you're stressed? Under the influence of stress hormones, the body mistakenly believes that you are in danger. It shunts blood flow from the gut to elsewhere in case you need, for example, to pump up your muscles to flee a burning house.

Darn that diarrhea! Ever wonder why some people experience "the runs" during stressful events or even in anticipation of stressful events? It's the body's stress response turning on to aid survival. If you had a full load of, shall we say, stuff in your colon that would keep you from running faster from that burning house, the natural stress response would be to quickly evacuate your colon before the mad dash. During the stress response, your upper gut muscles slow down—but the muscles of the large intestine go on a contraction spree, evacuating the gut of its contents before the water is absorbed.

~~~~~~~~~~~~~~~~~~~~~~~~~~

Some people are more sensitive to the gut effects of stress. Perhaps the term "sensitive gut" is also in the mind. Call it "sensitive mind gut," or "sensitive gut brain." (And turn to chapter five to learn how to not stress out both your gut brain and head brain.)

WHAT YOUR POO TELLS ABOUT YOU

Pooping is one of your gut brain's natural *detoxing* methods. Most of the dry weight of poop is aged-out gut bacteria that needs to pass on, or out, to enable the colon to repopulate with healthier microbes and keep the gut lining healthy. People who practice our healthy brain plan often report less bowel nuisances and discomforts, such as constipation and diarrhea. As a teen recently told Dr. Bill, "You were right! The better I chew the better I poo."

How well you poo is often a clue to your gut-brain health. Good gut health usually rewards us with at least three soft (but not watery) poos a day that slide out smoothly and taper to a tail at the end. (See AskDrSears.com/DrPoo for Dr. Bill's fun and informative read, *Dr. Poo*, about good gut health.)

7. MOVE MORE, MAKE YOUR MICROBIOME HEALTHIER

Why exercise helps with every aspect of your mental and
physical health: It makes your microbiome happy.
–Scott C. Anderson, coauthor of *The Psychobiotic Revolution*

In what we call the "Running Rats" study, microbiologists found that exercise made rats' microbiomes more diverse and increased secretion of *n-butyrate*, the healthy fat that nourishes the gut lining. When it comes to gut health, *diversity* is key, and one of the earliest findings of the microbe–health connection was that healthier people, especially those who eat the healthy brain diet we recommend in chapter two, tend to have a wider diversity of gut bugs. (See more about the movement–microbe connection in the next chapter, page 95.)

CHAPTER 4

~~~~~

MOVE SMART!

The more you move, the smarter, happier, and younger your brain. Doctors call exercise a smart "polypill" because it is good medicine for every organ, especially your brain.

What scientists say about how movement makes your brain healthier. In June 2017, Dr. Bill was invited to attend one of the best conferences on brain health. At this conference were not only many of the top neuroscientists in the world, but also more than a thousand healthcare providers and brain-health sufferers hungry for the latest medicines for brain health. Noted neurologist and author of *Brain Maker* Dr. David Perlmutter opened the conference with this theme: *"Exercise is the number one brain growth factor."*

It gets better. On Sunday afternoon we all gathered together for the closing keynote address, in which a top neuroscientist summarizes the top research. The announcement, from associate clinical professor of psychiatry at Harvard Medical School, Dr. John Ratey:

> *By increasing neurotransmitter activity, improving blood flow, and producing brain growth factors—Miracle-Gro or brain fertilizers—exercise*

readies our nerve cells to bind more easily and stronger. Exercise does this better than any other factor that we are aware of at the present time.

That's right: Movement is the best medicine for your brain.

WHY MOVEMENT IS SMART MEDICINE

Brain Needs	Movement Makes
• Better blood flow	• More blood flow
• Neurochemicals to grow brain tissue	• BGF and VEGF, neurochemicals that grow brain tissue
• Less neuroinflammation	• Inflammation balance
• More happy hormones	• Happy hormones serotonin and dopamine
• Growth hormone	• More growth hormones
• Smart sleep	• Sleep-better neurochemicals

Follow this logic. Your organs are only as healthy as the blood vessels supplying them. Remember, your brain is one of your top vascular organs, receiving 25 percent of the blood your heart pumps. Movement makes more blood flow to your brain. Therefore, movement makes your brain healthier. That was simple!

MOVE MORE, MAKE MORE BLOOD VESSELS

The more cerebral blood vessels, the better your brain. Suppose you were building a very busy city, but you wanted the traffic to move smoothly with no extra congestion. To make that happen, you would need to increase the number of roads throughout the city. When you move, you make better "roads"—blood vessels—within your brain. Also, suppose you want to ensure your city remained well fed. By increasing the number of roads, you make it easier for trucks to deliver nutrients more quickly. That's what more blood vessels do—deliver more nutrients to your needy brain.

Movement makes existing blood vessels healthier. It's not only the number of blood vessels that supply an organ that keeps it healthy, it's how wide open these vessels are. Sitters tend to have more narrow blood vessels, and the lining of their blood vessels tends to be sticky. And a sticky lining is prone to collecting clots.

Movement makes blood vessels more "fit." Just as exercise makes your muscles more "fit" (helps them perform better with less effort and exhaustion), movement also makes your vessels (which contain some muscle tissue) more fit. Fit vessels become trained to open wider more easily (with less blood pressure required), which promotes more blood flow to your brain garden.

MOVE MORE, MAKE MORE BRAIN-GARDEN FERTILIZER

Let's return to the garden analogy you learned in chapter one. Suppose you want to grow a fruitful garden. You realize that the garden can only grow if it gets enough water and fertilizer, so you create irrigation channels so water can reach each plant, and add the right fertilizer to help the plants blossom. The same works for your brain garden. Besides making more irrigation channels—blood vessels—as we just learned, movement makes natural brain fertilizers, neurochemicals called BGFs. These are known as the brain's natural wonder drug, and BGF levels tend to be lower than normal in the brains of patients with Alzheimer's. Just as plants wither without nutrient fertilizers, brain tissue withers when it's BGF deficient. Consider Alzheimer's disease the effect of worn-out brain tissue. One of my favorite other brain fertilizers is VEGF (*vascular endothelial growth factor*—sounds like "veggies"), a biochemical that helps keep the lining of blood vessels smooth.

MOVE MORE, MAKE MORE PERSONAL BRAIN MEDICINES

Besides blood vessels and brain growth fertilizer, movement makes a lot more brain medicines. Want to have some fun? Next time you're walking briskly with friends, proudly exclaim, "Oh, it feels so good to be making my own brain medicines!"* Naturally, the exercisers on either side of you will wonder what you're talking about. Tell them: "When you move, you make neurochemicals that make your brain smarter, healthier, and happier."

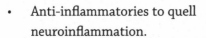

In addition to brain fertilizers BGF and VEGF, here are some other medicines movement helps your brain make:

I'm making my own medicines.

- Anti-inflammatories to quell neuroinflammation.
- Happy hormones (dopamine, serotonin, and more) that are your personal antidepressants and mood-mellowers.

OPENING YOUR BRAIN'S PERSONAL PHARMACY— A VISUAL GUIDE

The top theme of our healthy brain plan is how, by opening your personal brain-health pharmacy, you can make your own medicines, because the more medicines you make (by skills), the fewer you have to take (prescription pills). Naturally, you are wondering where the "personal pharmacy" is located in your body, what medicines it makes, and how you can open it (and keep it open) for your own brain health. Read on!

Where is your personal pharmacy? Suppose you were designing the smartest and most perfect machine ever made, the human body and

---

* *Make your own medicines*™ is the trademark teaching of The Dr. Sears Wellness Institute. At this writing we have taught MYOM to more than 10,000 students as certified health coaches and equipped them with the tools to teach others about this fundamental feature of brain health.

brain, but you knew that we humans were going to mess it up by how we think, eat, and live. Where would you put your personal health pharmacy? You would put it as close to your blood vessels as possible so these homemade medicines could be quickly dispensed into your bloodstream, to your brain, and throughout the rest of your body. That's exactly where this pharmacy is—in the lining of your blood vessels, called the *endothelium*, or what we refer to as your "silver lining."

How do I open my personal pharmacy? Through one simple change: "Move more!"

By now you should be thinking, "Wow! How does movement make my own medicines?" The answer to your question won the Nobel Prize.

A Nobel night with Dr. NO. Dr. Lou Ignarro and his wife, Sharon, were guests at Dr. Bill's home for his traditional salmon-at-sunset dinner. As they perked up their brains with the "smart fats" dinner, Dr. Ignarro taught Dr. Bill his Nobel Prize–winning research on how movement makes your own medicines.

As Dr. Bill was drawing the diagram you see on page 101— showing that within your endothelium, these medicines are housed in trillions of microscopic "medicine bottles" (they actually do look like miniature squirt bottles!)—Dr. Ignarro explained: "When you move, it makes your blood flow faster over these medicine bottles, creating an energy field called a *shear force* that opens your pharmacy and releases your personal medicines—beginning with nitric oxide, NO." (That's why Dr. Bill dubbed him "Dr. NO.")

Movement makes a NO-brainer. The main medicine your internal pharmacy makes more of when you move is *nitric oxide* (NO), which is really a multipurpose medicine. It dilates, or widens, blood vessels during exercise, like naturally widening a highway for more traffic to travel on during rush hour. Nitric oxide is also a neurotransmitter, one of those vital messengers that carry neurochemical messages from one nerve fiber to another and are vital to your ability to think smart and feel emotionally balanced.

Furthermore, NO is a natural antioxidant, or anti-rust neurochemical. As you learned on page 8, brains need lots of antioxidants because they are prone to rust. A new discovery shows that the brain also makes its own NO—a revelation that is prompting neurologists to wonder if Alzheimer's could be a NO deficiency. Since aging brains are more vulnerable to rust, aging brains need more NO. This is why brain health experts preach: "The older you are, the more you need to move."

~~~~~~~~~~~~~~~~~~~~~~~~~~~~~~~~~~~~~~~

Take a visual snapshot of the two images on the following page and plant them in your brain's visual-file library. Replay them often. You can also download a color version from DrSearsWellness.org/brainhealth. As a reminder to move more, paste it throughout your home and workplace, and in your cell phone gallery.

Notice that during exercise, the blood vessel of Mandi the Mover is wider and the blood is flowing faster. This creates the force that opens the medicine bottles in the endothelium and releases NO, which further dilates the arteries and improves blood flow. (The tiny bubbles in the medicine bottles are the molecules of *NO gas*, which is a smart design, since gases dissolve quickly in liquid blood. To help medical students remember Dr. Ignarro's Nobel Prize–winning discovery, Dr. Bill tells them: "Your endothelium passes gas.")

By contrast, notice how different Sam the Sitter's blood vessel is. Sam has two reasons why he's not making enough brain medicines. First, he puts *sticky stuff* in his mouth, which gets into his blood and accumulates on top of the medicine bottles, making them harder to open. Strike one! Also, his blood flows more slowly because he sits too much, so his medicine bottles remain closed. Strike two! Guess what strike three, sooner or later, will be?

VIEW IT, MOVE IT!

Visit AskDrSears.com/MYOM to see a Telly award-winning animated video narrated by Dr. Bill explaining all about how you "make your own medicines" in your endothelial pharmacy.

Mandi the mover and smart-food eater

Pharmacy open
No sticky stuff - medicines released

Sam the sitter and sticky-stuff eater

Pharmacy closed
Sticky stuff blocks release of medicines

MOVE MORE, MAKE MORE NATURAL MOOD MELLOWERS

Movers will be happy to know that there is a scientific basis to the wisdom in the statements "Walk it off" and "Movement mellows your mind." Short sessions of vigorous exercise are good medicine to calm you down from anxiety or lift you up from depression. "Runner's high" is a real thing! The mood-elevating and mood-mellowing feelings following intense exercise vary greatly among individuals, but people who do experience runner's high describe the feeling as "pure happiness," "elation," "a feeling of unity with one's self and nature," "inner harmony," "super energy," and "lessening of pain."

New research has uncovered the hormonal basis for how movement makes happy medicines. It's called *endogenous opioid release*. There is now a term for these neurochemicals: *endocannabinoids*. These natural opioids could explain why exercisers tend to suffer less depression.

> Exercise is possibly the strongest antidepressant we have.
> —Dr. Peter Bongiorno, author of *How Come They're Happy and I'm Not?: The Complete Natural Program for Healing Depression for Good*

BIGGER BELLY, SMALLER BRAIN

"The leaner your belly, the lower your mood disorders" is the mantra of *psychoneurogastroendocrinology*—the study of how your mind controls your hormones and how your belly size affects your mind. Abdominal fat is one of the main regulators that stabilizes insulin levels. In other words, the insulin level of your blood is related to the obesity level of your waist.

Suppose you entered our Brain Health Center for a check-up, but we had time for only one quick measurement that would most reveal your risk of getting a brain disease. Your *waist size* would be that telltale measurement.

How a big belly shrinks your brain. While diet and exercise are the two most familiar determinants of belly bigness, there is another habit that puts on pounds and inches around the waist—*stress.*

Your toxic belly fat is shrinking me!

Excess stress hormones, especially when they're too high for too long, release *cortisol*, a belly-fat storage neurohormone: it sends a text message to its fat-storage buddy, insulin, to store more belly fat. The big, fat belly cells then dump their excesses into the blood by spewing out sticky stuff, those neuroinflammatory chemicals that damage brain tissue and stick to the walls of your blood vessels. This slows blood flow to your brain. Slow blood flow = slow brain thinking. As Dr. Bill explains to his patients, "Your brain is getting foggy because your belly is getting saggy."

Being overfat, especially around the middle (we call it "toxic waist"), is a downward spiral. Excess abdominal fat prompts you to overeat because this type of fat secretes neurochemicals that travel to the hypothalamus and prompt the brain to eat more. *Increased abdominal fat promotes increased hunger.* Which tends to lead to further increases in abdominal fat.

Belly size and Alzheimer's. In his practice, Dr. Vince has seen a direct correlation between belly size and Alzheimer's disease (AD). Women with waist sizes of more than 35 inches have a fivefold increased risk for AD, and men with waist sizes greater than 40 inches have a threefold increased risk.

The connection between belly fat, insulin insensitivity, and Alzheimer's is so strong that in Dr. Bill's medical practice, he's upgraded the label "You're overweight" to the life-changing "You're *pre-diabetic*" or "You're *pre-Alzheimer's*."

The good news is our healthy brain plan is the way to get and stay "pre-."

MOVE MORE, AGE LESS

The older we are, the more we need to move.

Ready for some "wow!" research? One of the largest health studies ever done, the Nurses' Health Study, showed that those who exercised more became smarter and decreased their risk of Alzheimer's disease by two- to threefold. Other research has shown that older people with a life-long history of exercise are more likely to preserve their smarter brains.

Research reveals that exercise can help aging people develop a larger brain. After six months of aerobic exercise, study participants grew *larger brains*, both in gray and white matter. The researchers concluded that the increase in brain volume was most likely due to an increase in BGF, an increase in small blood vessel growth, and more interconnections between brain cells.

What's fascinating about this research is that the areas of the brain most likely to deteriorate as we get older—the prefrontal cortex, which enables quicker and smarter decision making, and the temporal lobe, which is associated with memory and other critical cognitive processes—were also the areas most improved by exercise.

The older we are, the more we need to move.

MOVEMENT HELPS HEAL THESE AGE-RELATED BRAIN CHANGES
You don't have to be one of these "old" statistics!

Our brains shrink. The aging brain can shrink from 3 pounds to 2.6 pounds in just fifteen years, beginning around age fifty. The primary areas that shrink are in the hippocampus and prefrontal cortex, the centers of memory and mental energy. Our healthy brain plan, and especially movement, helps keep the brain from shrinking. (Read more about brain shrinkage, page 18.)

Belly fat increases. As we age, our ability to eat carbs but not put on more belly fat (like some teenagers can do)—called "carb tolerance"—decreases. So not only do we need to eat less refined carbs, we need to build more muscle—through movement—to burn the carbs we do eat before they make their way to our bellies.

Blood flow slows. When we get older, we're at higher risk of developing something called *cerebral hypoperfusion*, or decreased blood flow through the brain. When blood flow slows from sitting too much, blood vessels get stiff, their lining gets sticky, and the brain's undernourished tissue shrinks. Movement increases blood flow to help prevent those brain S's: slow, stiff, sticky, and shrunken.

Vessels "wrinkle." With age, blood vessels start to lose their endothelium and so become more tortuous (the reason we say aging also causes "wrinkling on the inside"). Just as traffic flows more slowly on a curvy highway, so does blood flow slowly in twisted vessels. Some brain regions are more vulnerable than others to this drop in blood flow. The hippocampus and cerebral cortex are especially vulnerable.

When blood vessels lose their endothelium, they also lose the fertilizing VEGF that performs blood vessel maintenance and repair, like the maintenance crew on a highway. Our silver lining gradually wears out. And as our blood vessels' ability to repair themselves lessens, and the aging brain becomes less able to grow new vessels, blood flow and VEGF production slows further. Deprived of VEGF, brain tissue withers, like unfertilized and unwatered crops on a farm that eventually dry up and wither away.

The good news is this doesn't have to happen to you. Some people seventy-five years of age have similar vascular density to much younger people. Why? They move more and sit less.

Blood pressure rises. If aging vessels get stiff, the heart must pump harder to push enough blood through them. This causes blood pressure to be too high, which can further weaken blood vessels and sometimes lead to brain bleeds.

High blood pressure, or hypertension, causes increased rigidity of the blood vessels and shrinks their diameter, leading to decreased blood flow to the brain and often multiple mini-strokes. People with hypertension have a higher incidence of Alzheimer's and other neurodegenerative diseases.

Fortunately, our brain-health plan is just what the cardiologist would prescribe for lowering high blood pressure.

Growth hormone drops. Movement is your fountain of youth. One of the top youthful hormones in your brain is *human growth hormone*, a top brain-garden fertilizer. Sadly, the older we get, the less growth hormone we produce. It can decrease as much as 75 percent in men and 20 percent in women. Fortunately, exercise stimulates the production of more of these youthful, brain-fertilizing hormones.

The "sitting disease" is a downward spiral. The more you sit, the faster your cognitive decline, which can lead to inability to exercise, which increases cognitive decline further. In fact, just a one-mile walk a day can reduce your chance of early dementia by 50 percent. And the earlier you start, the better.

One of Dr. Bill's patients, an investment banker named Kirk, once defended himself: "But I don't have time for all that diet and exercise stuff. I have a bunch of lunch meetings and I work long hours."

"Kirk," Dr. Bill countered, "if you don't have time for diet and exercise, you better save more time to spend in the hospital and more money for doctors' bills. As a smart investor, naturally you have an IRA, right?"

"Of course!" he replied.

"You also need an IRAB—an individual retirement account for your brain."

Dr. Bill won his case and Kirk went home with a new retirement plan for brain health.

If you keep your silver lining healthy, you can better enjoy your silver years.

MOVE MORE, SLEEP BETTER, STAY SMARTER

The more you move during the day, the better you sleep that night. As you will learn in chapter six, daytime movement creates more "sleep prompts," neurochemicals that help you sleep better. And since smart sleep is good for brain growth and repair (see chapter six), that's just another reason to move more!

MAKE BEAUTIFUL MOVING-BRAIN MUSIC

It's not so much your "hormonal level" that determines health, but your *hormonal balance*. Exercise balances your hormones. There are more than 100 hormonal players in your brain's symphony orchestra, and you have previously learned that when you feed these players smart foods, they play better music. That's true! But exercise is another conductor that guides each player to play their best: not too loud (anxiety), not too soft (depression), not out of time (ADHD). When your brain's calm, happy, focused, and creatively thinking, and players are all working in harmony with one another, beautiful music—better brain health—is the result.

BEST MOVEMENTS FOR YOUR BRAIN HEALTH

The best exercise for you is the one you will consistently do. Enjoy it! When we enjoy our exercise, we get the brain-building effects without the brain-harming effects of stress hormones, which may increase when we are forced to do an exercise we don't want to.

Generally there are three types of exercises that are good for the brain. All three build smarter brains, but in different ways. Best for brain health is to do all three—strength building, aerobics, and isometrics—at least three times a week.

STRENGTH-BUILDING EXERCISES

These are especially smart for people over forty, for several reasons. Muscles are like brain cells—use them or lose them. Without exercise we begin losing muscle mass around age thirty. Then we tend to lose 10 percent of our muscle mass every ten years after age fifty. When you lose muscle, several brain-unhealthy effects occur. Since muscle, like the brain, is a big sugar user, when you lose muscle you lose a sugar balancer. The more muscle you have, the more likely you are to maintain insulin sensitivity—meaning that your cells use the glucose you consume more efficiently—and avoid sugar spikes. Remember that bad word for brain health: spikes.

Perhaps the biggest brain-health effect of strength training is that this type of exercise stimulates your body to make more *growth hormone*. When you heavily exercise a muscle to its fatigue point—in gym-rat lingo,

We're growing our brain gardens.

"feel the burn"—you exhaust your mitochondria, those millions of tiny energy batteries within each cell. (See more about your mighty mitochondria, page 6.) These energy-needy mitochondria quickly send a biochemical text message to the brain, telling it to make more human growth hormone, which builds more mitochondria and reenergizes the muscle. When this hormone sends a message to build more muscle, it also sends a message to produce more BGF, which builds more brain tissue.

MOVE MORE, FALL LESS

Preventing falls is another reason why exercise, and especially strength-building exercise, is the fountain of youth: falls are one of the top causes of senior disabilities.

Exercise increases muscle, and the stronger your muscles, the less you fall. The less you fall, the fewer disabilities you suffer and the better your brain. Also, the stronger your muscles, the stronger your bones have to grow to support them. The stronger your bones, the less likely they are to fracture or break if you do fall.

As Dr. Vince likes to say, "An hour a day of movement may keep you from later spending twelve hours a day in a wheelchair."

TREES TRUMP TREADMILLS

To get the most brain-building bang for your walking buck, movement in nature is best. Why? Walking on uneven surfaces— grass, sand, twigs, and so on—requires more muscles. And, quite frankly, the brain gets bored moving on a monotonous treadmill. (See more about moving in nature, page 113.)

THE HEALTHY BRAIN BOOK

AEROBICS

Aerobics, or whole-body movement that gets your heart pumping (also called "cardio" or cardiovascular), includes walking, running, swimming, and dancing. Aerobics mainly benefit your brain by increasing blood flow. As you learned on page 96, increased blood flow to your brain releases brain-fertilizing and mood-mellowing hormones.

Remember the study that showed aerobic exercise helped aging people develop larger brains? It also showed the benefits of aerobic exercise for brain growth over other forms of exercise. The fifty-nine healthy older adults in the study were divided into two groups. One group did lots of cardiovascular exercise, also called aerobic exercise; the only change in the other group was they did nonaerobic exercise, or primarily stretching and toning exercise. After six months, follow-up MRI neuroimaging studies showed that while those who participated in aerobic exercise grew *larger brains* both in gray and white matter, those who did only stretching and toning did not.

HIIT YOUR BRAIN HEALTH

High-intensity interval training (HIIT) is a favorite of trainers and a time-efficient way to get more bang for your exercise buck. HIIT means blending longer periods of moderate exercise with periodic spurts of more intense movements. For example, while walking, every five to ten minutes vigorously walk faster for a minute, or long enough to feel your heart beating harder and your breathing getting heavier.

HIIT helps you provide a higher level of brain-garden fertilizers in less time. And Mayo Clinic researchers who studied how HIIT helps overall fitness concluded that this type of exercise triggers cellular DNA to produce more energy, mostly by making those mighty mitochondria you learned about on page 6 more energy efficient.

Another HIIT highlight. People who need HIIT the most, profit the most. Mitochondrial function tends to get weaker as we get older, yet research shows that seniors get a greater mitochondria boost from HIIT than do younger exercisers.

Need a lift? Take a HIIT. To quickly lift out of a funk, and when depressed, performing a 30- to 60-second HIIT exercise will lift your spirits. Dr. Bill takes morning HIITs. During his morning walk or elliptical workout, adding a few HIITs perks up his brain and his mood to think and feel better.

How to HIIT. As a general guide, begin with a 5:1 easy-to-hard ratio, such as five minutes of moderate walking to one minute of a really fast pace. Gradually progress to a 3:1 routine. To give your body a rest, you may wish to limit HIIT exercises to half of your weekly workout, such as twice a week.

Whenever you feel down or your brain needs a lift to be happier and more creative—take a HIIT!

ISOMETRICS: THE ANYTIME, ANYWHERE SMART EXERCISE THAT ANYONE CAN DO

The best exercise for your brain-health increases blood flow to your brain *and* makes more brain-growth fertilizer. Isometrics does both.

Isometrics are a muscle-maintaining and muscle-building tip Dr. Vince learned as a member of the Olympic weightlifting team. They're *anytime, anywhere* exercises because your bones don't move; only your muscles flex.

In isometrics, you simply flex large muscles, like quads, biceps, glutes, and abdominal muscles (when possible, pressed against other muscles or against a table). Then you hold them for 30 to 60 seconds until you feel the burn. (Remember, when you feel the burn, so do your mitochondria.)

Isometrics are especially helpful for sitters who spend a lot of time in "boring" meetings. You can be lifting and holding your legs for a minute or two under the table, and no one will know. You can put your laptop screen in front of you while you flex your biceps and pecs. You'd be surprised at the end of an hour or two-hour meeting how your body feels like it's had a good workout. If you're a SOW (sedentary office worker), isometrics are especially for you. (See "Endnotes" for more brain-health perks from isometrics.)

> See Drs. Vince and Bill demonstrating isometrics at
> AskDrSears.com/isometrics.

THE SMARTEST EXERCISE?

Imagine once upon a time a bunch of brains got together to discuss their favorite exercise. One brain chose golf: "Movement in nature is so peaceful." All agreed. "I like weightlifting," a hunky brain boasted. He got only a few "likes." A mellow brain chose swimming: "Those floaty feelings not only give me a good workout but a relaxed mind." A brain at a desk job all day boasted about his isometrics workouts while sitting. The head brain put all the voters together and decided they were all right. The resistance exercises in strength training boost brain growth hormones; vigorous movement increases blood flow to nourish our brain garden.

For those who like an all-in-one movement machine, an *elliptical trainer* is smartest because it combines cardio and resistance exercises. Dr. Bill's elliptical is the star player in his home gym, or what his grandchildren call "grandpa's playroom." See Dr. Bill demonstrating elliptical exercises at AskDrSears.com/MYOM.

MOVING IN NATURE: EXERCISE SQUARED

Remember Dr. Mom's prescription for boredom and bad behavior? "Go outside and play!" No neuroscientist has ever topped this advice from Dr. Mom. Modern brain scans, those windows into your brain, reveal she was right. In our medical practices we often use phrases like, "You're suffering from the *indoor disease*" or "You have NDD—*nature deficit disorder*."

"Nature is the best medicine," one of the oldest and wisest pieces of advice, is even more fitting in today's artificially stimulated and over-loaded society. Nature neuroscientists teach that once we started working and playing indoors instead of out, we started losing our minds.

More movement while spending time outdoors is good brain medicine. Movement in nature, we believe, reminds your brain, "This is how you were genetically wired thousands of years ago. Welcome back!" Sitting indoors in windowless rooms staring at artificial light is a mis-fit for the mind. (See related sections: "Eye Feel Good," page xvii; and "Enjoy Moving Meditation," page 153.)

As neuroscientists Eva Selhub and Alan Logan mention in their book *Your Brain on Nature*, "Green exercise is like exercise squared."

BRAIN-HEALING BENEFITS OF NATURE THERAPY

When we take a brisk walk in the woods, a new medical specialty, the neuroscience of nature, has proven that we:

- make more happy hormones (serotonin, dopamine).
- mellow our moods.
- lower our stress hormones (cortisol).
- lower our heart rate.
- lower high blood pressure.
- lessen neuroinflammation.
- prevent and delay dementia.
- create smart ideas.

- think happier thoughts.
- feel more "positive."

A PEEK INSIDE YOUR BRAIN IN NATURE

Suppose you know you feel good when you enjoy the great outdoors, but you want to know why. So, on your way to the park, you stop by your friendly neighborhood neurologist. At the appointment desk, you announce, "I'm here to get wired." The receptionist smiles and escorts you to a room where you get fitted with a cap full of wires that is remotely connected to an MRI scanner that is going to record what's going on in your brain while you walk in the woods.

As you begin your walk, you smell the roses, look at the trees, and take deep breaths to smell the leaves. You feel good. Meanwhile, back at the neurologist's office, the MRI scanner is showing why. Areas in your brain, called the *parahippocampal gyrus*, release natural happy hormones, opiates and dopamine. As you take in more delights of nature, the MRI shows your brain dialing up the relaxation centers of your brain and dialing down the hyper, or anxious, centers of your brain, also called the arousal centers—the parts of your brain that get revved up during rush-hour traffic.

As you continue your walk, you notice the *nature effect*: your mental fatigue, or brain fog, is much less while viewing the scenes of nature than during your morning commute. You're feeling more "positive." You like yourself more and appreciate what you have. You leave the park feeling happier and healthier, both physically and mentally.

Go outside and play, lower your stress. Measurements of stress hormones in body fluids (saliva, urine, and blood) show that stress hormones drop when we move outside, too. In a study done by the Department of Forest Medicine of Nippon Medical School in Japan, levels of adrenaline in the urine of participants who spent a day of walking in the woods decreased by nearly 30 percent. The stress hormone levels dropped even more in the women in the study than they did in the men.

Go outside and play, make more DHEA.
Neuroscientists have also discovered that spending time moving outside increases a hormone called DHEA, which helps maintain the level of brain-growth fertilizer. Unfortunately, DHEA can decrease as we age—if we let it. The older we get, the higher the dose of go-outside-and-play we need.

Move me more ...outdoors.

GO OUTSIDE AND PLAY: "RE-CREATE"

During a walk in the woods, Dr. Bill had an "aha!" realization. *Recreate* should mean getting involved in an activity that *reminds* the brain (re-minds) where it feels and thinks best: the *outdoors*. A "recreation center" with a basketball court, swimming pool, and ping-pong tables can be good indoor medicine for the mind, especially in nasty weather. But to really "re-create," go outside and play.

TIRED WHILE MULTITASKING?

Ever notice how your brain gets tired and foggy during multitasking? Back to the "brain as symphony orchestra" analogy. Ideally, the entire orchestra, and its conductor, is focused on playing the piece in front of them—say, the "Blue Danube" waltz. Beautiful music results. Suppose, however, that a few players let their attention wander to another piece. Disharmony results.

Neuroscientists call this split focus *attention fatigue*. Trying to process too much at once results in distraction and mental exhaustion from our efforts to focus despite that distraction. Recognizing this new *attention fatigue disorder*, neuroscientists came up with a solution to the problem, which they refer to as "cognitive restoration" or "attention restoration therapy" (ART). Translation: Take a break, go outside and play!

TAKE A THINK-WALK

You've heard the sage advice, when you're facing a tough problem, to "sleep on it." Try "walking on it" instead! Thoughts come together, distractions tend to disappear, and mental clarity blossoms during a walk. Double that for walking in nature.

Much of this book was written while walking, each chapter dictated into a recorder; Dr. Bill's current average is one-and-a-half to two hours a day.

MOTIVATION TO MOVE

As we've seen, movement is good medicine for your mind and body. If you were to do a literature search on the medical benefits of movement in nature, you would discover they exceed any store-bought drug:

- Lessens stress
- Releases happy hormones
- Reduces the pain threshold
- Grows a bigger and smarter brain
- Makes the *antis*: *anti*depressants, *anti*-anxiety, *anti*-inflammatories, and *anti*oxidants
- Helps heart health
- Improves lung function
- Facilitates digestion
- Eases arthritis
- Lessens type II diabetes
- Reduces inflammation

Even knowing all these benefits, many people find it hard to get moving. Even when our patients know intellectually that they need to move more, they still often say things like, "But I just can't make myself do it" or "I've been meaning to do it." It can be even harder for depressed people, because depression dials down self-motivation. How to pull the sitter off the couch has been a top challenge in our medical practices. Here's what we've found works:

MOVEMENT RULES FOR BRAIN HEALTH

1. *Move green.* Walking outside, preferably in parks, forests, golf courses, and so on, is most likely to generate the quickest feel-good effect.
2. *Move wet.* (See flotation therapy, page 154.)
3. *Move fast* to feel better fast. (See HIIT, page 110.)
4. *Move together.* Having an exercise buddy keeps you both accountable.
5. *Move techie style.* Your exercise buddy can challenge you: "How many steps did your step counter record yesterday?"
6. *Visualize what smart medicines you are making in your body and brain.* Ponder the mover-versus-sitter figures on page 101. Imagine: "That's really going on in my body and brain. My beautiful pharmacy inside is making just what I need."

Our closing "prescription" for brain health:

Go outside and play!

CHAPTER 5

THINK SMART: EIGHT
STRESSBUSTERS
TO THINK-CHANGE
YOUR BRAIN

While no neuroscientist has been able to trump Dr. Mom's stressbuster "Go outside and play," in this chapter we will add more of them to your toolbox, starting with:

> **You can think-change your brain.**

You are what you think. More correctly, you are what you let yourself think. In the earlier chapters you learned how you can build a better brain by changing your diet and exercise habits. Now we add another brain-changer: how you can change your brain by changing your thoughts.

When we say "stressbusters," we really mean *stress balancers*. Being completely without stress is a near impossibility living in the real world. We want you to have "smart stress," meaning you are able to dial up your stress neurochemicals when needed and dial them down again afterward, when they're not needed at such a high level. Stressbusting means knowing and feeling when to heat up and when to cool down—when to express strong feelings and when to just chill.

In our medical practices, when teaching patients to develop their own toolboxes of personal stressbusters, we summarize it simply: "We want you to learn how to *self-regulate*."

In delving into the mind science of *psychoneuroimmunology*, you will learn that it's the extremes that make you sick and tired. People who are quick to "fly off the handle" are at higher risk of becoming chronically anxious. On the other extreme, people who are *emotionally flat* are at risk of illnesses affecting their immune system, such as cancer and autoimmune diseases.

As we give you a crash course in stress management, we want to take you back into your brain garden. On one side are flowers that bloom beautifully all year long, providing you fertilize them. Happy, positive thoughts increase the blood flow containing fertilizing neurochemicals, such as dopamine and serotonin. The flowers grow and eventually take over the garden and you become a "positive person," a "happy person," because you've grown your happy center in your brain garden.

Yet suppose you let unmanaged stress and negative thoughts (toxic thoughts), like weeds, take over your garden. Then the flowers aren't fertilized enough to crowd out the weeds. So, you're left with a weed garden, and become a negative, unhappy, pessimistic person. You're about to learn how to fertilize your flowers to crowd out the weeds.

First, let's take a look at how stress affects your brain.

YOUR BRAIN NEEDS BALANCE: THE NEUROCHEMISTRY OF CALMNESS

In your brain, there are two systems that control your actions. One is very easy to understand; let's call it the *voluntary nervous system*. When your thoughts give a command, say, to scratch your ear, this system quickly sends a nerve impulse through a neuroelectrical circuit that moves your arm and fingers and gets right to the ear all within seconds. Then the brain tells your finger to stop scratching.

The second system is called the *autonomic nervous system* (ANS). It acts as an internal regulating center, automatically adjusting many bodily survival functions, especially our heart rate and blood pressure.

The ANS is divided into two subsystems, the sympathetic nervous system (SNS) and the parasympathetic nervous system (PNS). The SNS is the turn-on or "rev up" system. It kicks into action whenever you experience stress—when your brain perceives that there is an emergency. It puts you on high alert, helping you to be more vigilant and react faster, by instructing your nervous system to release adrenaline and its "be more alert and vigilant" buddy, noradrenaline. Your SNS turns on whenever someone, even jokingly, sneaks up behind you and grabs you. It is also turned on during sex.

The PNS, or "cool it" system, plays an opposing and balancing role: it calms you down and helps you sleep. It's also called the "rest and digest" system, because it promotes growth and energy storage. It's what clicks in to give you those after-meal drowsy, satisfied feelings.

Most of our day is spent turning these systems up and down to react and adapt to changing situations. The SNS speeds up the heart; the PNS slows it down. The SNS diverts blood to your muscles when needed to run from a burning building; the PNS diverts blood flow back to your digestive organs once you've made it to safety. Life changes and your brain is set to adapt.

SNS	PNS	Effects
Dilates pupils	Constricts pupils	Adjusts light to facilitate vision
Accelerates heartbeat	Slows heartbeat	Adjusts blood flow
Inhibits salivation	Stimulates salivation	Regulates salivation
Slows digestion	Stimulates digestion	Regulates digestion

SNS	PNS	**Effects**
Stimulates epinephrine	Turns down epinephrine	Revs you up or calms you down
Stimulates sexual arousal and orgasm	Also stimulates sexual arousal	Facilitates procreation

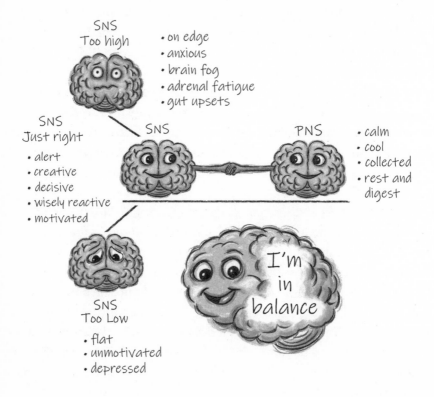

The ANS includes an important built-in safety feature: activating one of these systems automatically deactivates the other. Think of the SNS as the gas pedal and the PNS as the brake. The brain makes sure you can't press both at the same time. Most mood disorders are due to an imbalance: pressing the gas pedal too hard for too long (anxiety), or putting on the brakes too soon and too often (depression).

When your SNS and PNS dials are in balance, the body is in balance. But when stress sets the SNS dial too high for too long, the constant anxiety literally depletes your brain of energy, causing brain fog.

The adrenaline you release turns on cortisol, an excess of which literally causes your brain to become "worn out." Yet, if your SNS remains set too low, you become depressed, unmotivated, and have a "flat" mood. Being able to dial up your SNS to be creative, think quickly, perform, or avoid danger, and then being able to "cool it" and dial up the PNS to put the brakes on the SNS, is key to a healthy, balanced brain. Our healthy brain plan helps optimize this balancing act.

SLEEP STRESS AWAY

When you're asleep, your stress hormone dial is, too; it's turned down to its lowest level. In fact, we experience what's called a *circadian rhythm* when our natural stress hormones are at their highest level in the morning. Anthropologists believe this is because stress hormones needed to be highest when humans most needed to be on alert for hungry animals searching for them. Our stress hormones naturally dial down to reach their lowest levels just before nightfall and then gradually dial back up to be high again around sunrise.

SMART STRESS, DUMB STRESS

Let's take a closer look at when stress is smart, and when it is not.

Your brain and body on smart stress. Suppose the fire alarm goes off. You awaken, and your stress hormones dial up to high alert. Why? Because the well-being of your family depends upon your making quick, life-saving decisions. Your SNS tells your blood pressure and heart rate to increase to pump more blood to your brain, heart, lungs, and muscles while shunting blood away from your gut, kidneys, and reproductive organs. You don't have time to eat, pee, or procreate while your house is burning. You grab the kids and, if time permits, your prized possessions, and you safely run out of the house, escaping the fire.

You take a deep breath, think something like, "Thank God we're all safe," and your stress hormones dial back down. The next day you get on with the repairs, still thinking about how fortunate you are to have survived unharmed.

This is a normal and healthy stress response.

Not smart stress. Suppose, however, that after you escape the fire, your stress dial stays high. You continue to worry: "If only I had . . . " "What if it happens again?" "Why me?" The worry drags on for days and weeks. Your stress hormone dial stays up and the stress effects of high blood pressure and high heart rate stay high. When stress hormones stay too high for too long, instead of helping the brain, they damage it. This process is called *glucocorticoid neurotoxicity* (GCN). Translation: "Hey, high stress hormones, you're wearing me out!"

COACH HAYDEN NOTES . . .

One way to help get your brain back to a normal stress state is to acknowledge, possibly even out loud, that your high stress response did an amazing job "saving the day" and needs a well-deserved rest. It was there for a reason but is not currently needed, and you can take it from here. This simple act of acknowledgment can go a long way to helping you de-stress and rest.

YOUR SKILLS ARE BETTER THAN PILLS

One of the most important goals of our healthy brain plan is keeping these stress dials in working order: not set too high or too low; turning up and down when they should. Some medications help do this, such as turning down the high-stress dial for people who are manic and anxious, or turning up the dial for people who are depressed.

The problem is, unlike the medications you make yourself, the medications you take seldom get it right. They either turn the dial down too low, so that you become "flat," or they turn the dial up too high, so that you become hyper. This imbalance is especially evident in people who are on three to five brain-changing medicines; we call it the *cocktail effect*. Unlike the players in the body's natural symphony orchestra, these drug players don't communicate with one another. Your natural mood-mellowing neurochemicals are all interdependent. When one plays too loud, the others play differently to compensate, or the conductor gives them a "soften the noise" gesture. The drug players all do their own thing—playing too loud, too soft, or too long. Disharmony may result!

DON'T WORRY, BE HAPPY: HOW STRESS BOTHERS YOUR BRAIN

Stress hormones are somewhat like blood sugar. Our brain needs just the right level to function: too low and the brain gets too tired; too high and the brain gets inflamed and damaged by sticky stuff. The same is true for stress hormones. Too high and it can cause inflammation; too low and it can lead to depression. The right level is biochemically good for the brain.

Stress hormones go by various names, such as cortisol, corticosteroids, glucocorticoids, epinephrine, norepinephrine, and many others. We'll mainly discuss the stress hormone *cortisol*, because it's the most common and best known, but know that it's really not accurate to blame all of our stress responses, and their consequences in the brain, on this one hormone.

How worry shrinks the brain. Prolonged high levels of cortisol can literally shrink the memory and emotional centers of your brain. This is why the worst thing you can do when going into an important presentation is to be overstressed, because it can neurochemically shut off the memory bank that you need to draw on to give it your best. While the right level of cortisol increases alertness and attention during a work task or while taking a test, stress hormones that are too high for too long dull focus. Your mind can "go blank." It's that glucocorticoid neurotoxicity again.

Why stress makes you tired. Wonder why stressed and anxious people often suffer "chronic fatigue"? When you experience a stressful situation, your lifesaving stress hormones (especially the fast-acting adrenaline rush) quickly withdraw fuel energy from your body's energy storage bank (stored sugar and sometimes fat) and deliver it to the brain and muscles for quick decisions and fast movement. But once your stress dial turns back down, those hormones stop withdrawing the energy, and your body has the chance to build its energy stores back up.

When this stress dial stays turned up too high for too long, continuously calling on its buddy, the adrenal gland, to produce more cortisol, your energy bank eventually goes broke. You get sick and tired, something called *adrenal fatigue*.

MARTHA SEARS'S STRESSFUL STORY

While writing this book, Dr. Bill's extended family suffered the highest doses of stress they'd faced in decades: cancer, depression, anxiety, divorce, and more. His wonderful wife Martha, usually the queen of stress management, crashed. Her mind-set went from "Don't worry, be happy; that's a smallie" and other self-help prompts you'll read about later in this chapter, to being constantly "on edge," quick to fly off the handle, and experiencing other stress effects. Her gut brain got the brunt of her head-brain imbalance. One night her intestines literally shut down. Her floppy colon twisted on itself and became obstructed, requiring emergency surgery to relieve. We doctors, and husbands, are learning more all the time about how many

---

modern illnesses are triggered by stress. (See Endnotes to read Martha's amazing and motivating testimony on how she used our healthy brain plan to transform from debilitating depression to a happy, vibrant person.)

HOW CHRONIC STRESS CAUSES CHRONIC DISEASES

Many serious illnesses begin with stress in the brain, especially diseases of the heart, gut, thyroid, and immune system. Here's a head-to-toe story of how stress management is self-care for the brain and body.

Excess stress releases sticky stuff. During a normal stress response, biochemicals such as cortisol, epinephrine, and glucagon raise blood sugar levels, which gives the body more fuel. If your blood sugar stays high for longer than you need it, instead of being used by your muscles, the extra gets stored as fat or gets "glycated" (sticky) and deposited onto tissue as AGEs (advanced glycation end products)—biochemical jargon for gumming up the workings of your body and brain.

Excess stress causes a neurotic heart. Chronic unresolved stress increases blood pressure and heart rate and narrows coronary blood vessels. Eventually, these stress-related changes, especially high blood pressure, can wear out your heart.

The higher the pressure of the blood flowing through your vessels, the higher the force it exerts on the lining—sort of like how heavy trucks exert more wear and tear on the road. This can cause tears, or potholes, where sticky stuff collects more easily, setting you up for cardiovascular disease. Your immune system interprets this sticky stuff as foreign material and mobilizes the repair troops to "fix it." The problem is it's often not fixed and sometimes over-repaired, leading to more sticky-stuff buildup and clots, hallmarks of atherosclerosis (thickened arterial walls with bumpy plaques on the lining).

More stress-causing sticky stuff. During the stress response, epineph-rine makes circulating platelets more likely to clump together. These clumps of sticky stuff are more likely to collect in your blood vessels, especially around tears in the lining from high blood pressure, and cause atherosclerosis.

Excess stress causes pains in the gut. See the relationship between stress and gut health, page 92.

Excess stress weakens your immune system. Ever wonder why you get sick when you feel down? Because your head brain and gut brain (where much of your immune system resides) are so interdependent, what one feels affects the other. The neurochemicals from unresolved emotional stress don't stay in the brain. They get into the bloodstream and travel throughout your body, where they dial down the immune system, mak-ing it less smart. Sometimes it fails to attack an invading germ and you get a cold. Sometimes it gets confused and attacks your body's own tissues—as happens in autoimmune diseases. This is also called

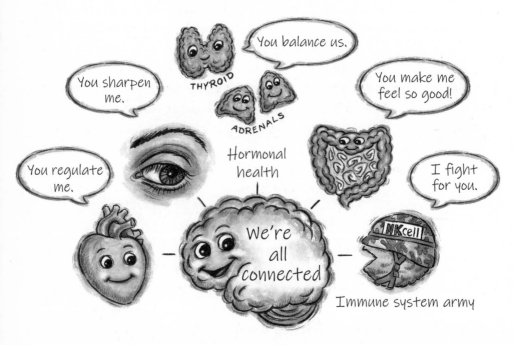

inflammation imbalance: the immune system remains *inflamed*, continuing to fight even after the battle is won, which causes wear and tear on the body.

Besides making antibodies against germs, your immune system helps you heal from illnesses. Because stress redirects energy away from the immune system, mobilizing it to help you handle stress, it deprives your body of what it needs to heal.

Continued high stress hormone levels are one reason chronically anxious people suffer more autoimmune illnesses. Remember, keeping your brain in balance helps keep your immune system in balance.

Excess stress can contribute to cancer. One of the other things the immune system does is search and destroy tumor cells. When stress suppresses the size of the immune system army, lowering the number of killer cells, it's just the opposite of what the body needs to prevent and fight cancer. And tumors love to be fed by the stress response's high levels of circulating glucose. This may help explain why "cancer fighters" who take charge of their disease and are optimistic about beating it more often do so, compared to those who are depressed and negative and give up. (See related section, "How the Belief Effect Benefits Your Brain," page 156.)

Excess stress makes you physically weak. The immune system isn't the only thing stress weakens—it weakens your physical body, too. Chronic high levels of stress hormones can break down muscle protein (stress hormones are looking for any tissues to extract energy from), leading to muscle fatigue; hence the current epidemic of feeling tired and in pain due to *fibromyalgia*.

THE OLDER YOU ARE, THE LESS STRESS YOU CAN HANDLE

Stress expert Dr. Robert Sapolsky defines aging as "the progressive loss of the ability to deal with stress." It takes less stress to suppress the immune system of an aging person than a younger one.

Many seniors also struggle with dialing down their stress response after a crisis has passed. It takes longer for their levels to return to baseline, and their resting levels of stress hormones are higher, too. This particular quirk of aging is thought to contribute to many of the common diseases of aging, including cardiovascular and cerebrovascular disease.

How high stress infects healthy aging also fits with the wear-and-tear theory of why we age. The less wear and tear on your body and mind throughout life, the longer and healthier you live.

Putting on your brain brakes. Most of our stress-control mechanism is located in the hippocampus, our memory center and one of the areas of the brain that loses the most neurons during aging. As we learned, the PNS system acts like a set of brakes. When stress hormone levels are high enough and the stressor, the reason for raising them, is over, the brain says, "Enough already, dial down!" and applies the brakes. With aging, the brakes in the hippocampus wear out a bit—and, unfortunately, they're not as easy to replace as the brakes in your car.

Because the ability to handle stress lessens as we get older, it's important to carry a larger toolbox of stressbusters as we age. Just as we must eat less as we age, we must also stress less. Our wish for our readers is that they start filling this toolbox with stressbusters at a younger age, which you will now learn how to do.

YOU CAN CHANGE YOUR BRAIN

We can change our brains—for better or worse—by changing our thoughts. The power of positive thinking now has scientific proof. We call this process *think-changing your brain*.

Pioneer think-changing studies—especially by Dr. Jeffrey Schwartz, research professor of psychiatry at UCLA School of Medicine, using PET scans of the brain—showed that changing habitual negative thoughts to habitual positive thoughts, through a therapy called cognitive behavioral therapy (CBT), can actually change the structure of the areas of the brain where thoughts occur. In other words, there are "happy" and "sad" centers in your brain—and you can affect whether and how much they grow.

You can think-change your brain

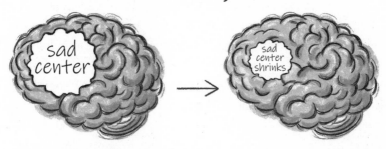

The left side of the above illustration shows an overactive area in the brain of a patient with OCD from one of Dr. Schwartz's studies. The area that's lit up is the "worry circuit," the prefrontal cortex. The image on the right shows the brain of the same patient after ten weeks of CBT, and with no medication. There's a clear decrease in size in the area of the brain that was previously "infected" with negative thoughts. By learning to convince themself that "This is really not me, it's a neuronal glitch. I don't have to pay attention to it," the patient was able to actually shrink the worry center in their brain.

These "wow!" studies opened up a whole world of new treatments based on personal thought power, or how the mind (thoughts) can change the structure of the brain. That is what this chapter will teach you how to do in your own brain.

YOUR BELIEF EFFECT

How well think-changing your brain works depends upon how much you believe in your own power to change your brain. Your *belief power* ("I believe I will heal") and its mental exercise partner *willpower* ("I'm determined to get my act together") can willfully change the sad centers in your brain to happy centers, as you just read.

While we don't yet have the technology to measure exactly what this "mental force" is, we believe your mind sends neuro-chemical text messages to the areas of your brain that are out of balance, in effect saying, "I believe my body-mind will heal, so heal!"

We doctors have long witnessed that those people who strongly believe they will heal, especially from illnesses like cancer and inflammatory diseases, are more likely to heal. This belief effect is also very powerful for imbalances in brain neurochemistry. (See related section on the belief effect in healing from cancer, page 156.)

HOW TO THINK-CHANGE YOUR BRAIN FROM THAT OF A NEGATIVE TO A POSITIVE PERSON

You've heard the expression "you are what you eat." This nutritional wisdom is true for the body and brain. What is also true for the brain is "you are what you think." You are the sum total of the thoughts, memories, and images stored in your brain. When you learn to dial down the negative and dial up the positive, eventually your brain becomes full of more positive images and you become a "positive person"—someone who is happier, healthier, and more enjoyable to be around. Yes, we are what we think.

Your brain's personal thought pharmacy. Just as the cardiovascular system boasts your body's largest natural pharmacy (see page 98), your brain has a pharmacy of its own. What opens this pharmacy to dispense its own brain medicines are mainly thoughts. Happy thoughts dispense happy medicines, like serotonin; unhappy and worried thoughts close your happy-medicine pharmacy.

The benefits of positivity. Doctors who specialize in preventive medicine have long known that people with positive outlooks on life tend to live longer and healthier lives. *Positivity promotes longevity.* Negativity promotes neurotoxicity. Makes sense! Positive people are better able to self-regulate with stress management.

Here's how positivity and negativity both affect the body:

Harmful Stress Effects of Negativity	Healthful Brain Effects of Positivity
• Dials up SNS too high	• PNS predominates
• Raises heart rate	• Lowers resting heart rate
• Raises blood pressure	• Relaxes stiff arteries
• Disturbs sleep	• Helps you sleep better
• Weakens immune system	• Balances immune system
• Promotes neuroinflammation	• Dials down neuroinflammation
• Leads to depression and/or anxiety (stress hormones too low or too high)	• Raises happy hormones
• Shortens lifespan*	• Promotes longevity*

* See "Endnotes" to learn how being positive lengthens your telomeres, part of your chromosomes, which can in turn lengthen your life.

LOOKING FOR THE POSITIVE

In Dr. Bill's frequent lectures on brain health, he purposely picks out people in the audience he calls "affirmers," those who are visibly affirming what he is saying. Their smiles, approving nods, and other polite gestures help him enjoy a more positive lecture attitude. One night after giving a lecture to a church audience, he thanked a few guests in particular for being nice "amen-ers."

A SMART TRIP THROUGH THE THOUGHT-CONTROL CENTERS IN YOUR BRAIN

Note to the reader: In our trip through the brain, we assign certain thoughts and emotions to certain areas in the brain, but scientifically it's not always quite that simple. So many areas are interconnected and synergistic that many areas may be involved in the process of think-changing your brain, in addition to the ones mentioned here.

To explain how think-changing your brain actually works, let's take a quick side trip inside your head. The *thalamus* is like the brain's air-traffic controller. All the outside signals you take in pass through the thalamus and get rerouted, hopefully to land on the right brain runway at the right time. The thalamus then signals the *hypothalamus*, the emotional pharmacy of your brain, to create a neurochemical response (positive or negative) to these thoughts.

The *amygdala* is the structure that decides whether a thought is good for the brain and body and should be filed in its library, or else quickly discarded into the trash bin. A hyped-up or anxious amygdala can act like the brain's "panic button" and trigger anxious thinking. Or the amygdala can "cool" its panic response and reroute it. The quicker a toxic thought is rerouted into the trash bin, the more likely it will be temporary, disappear, and not become part of your brain's file library. If it lingers, however, it flows into your *hippocampus*, which acts like a thought clearinghouse: when we allow a toxic thought to fester, the hippocampus decides it must be important, and files a photocopy of it into permanent or short-term memory, rather than trashing it. (The hippocampus does a lot of this work while we sleep.)

Here is where stress management becomes important. The hippocampus is one of the sections of the brain most vulnerable to unresolved stress because it has a lot of extra stress receptors. When a brain structure has "more receptors" (microscopic doors that let in neurochemicals), neuroscientists believe that this makes a particular structure *more sensitive* to the neurochemicals triggered by your thoughts. Think of "more receptors" as more buttons to push.

Think carefully! Here's the good news: you can choose whether your thoughts get sent to the brain's trash bin or stored long term. When you put a lot of energy into a thought, it's sort of like choosing to make a thought boldface instead of regular type, or writing it in indelible ink instead of easily erasable pencil.

New insights into the neurochemistry of thought suggest that the "harder" you think about a thought, the more neurochemicals you produce, and the quicker and deeper the thought is stored in your memory bank—whether you want it to be or not. Yet if you quickly trash a thought as soon as it enters your memory bank, the less likely it is to be deeply stored—the less permanent and more erasable the ink on your memory photocopy will be. Your goal is to quickly and easily erase toxic thoughts, and file happy thoughts in permanent ink.

It's your thoughts that build your memory bank, and therefore your brain. You want to mobilize three important brain areas whenever a thought starts to enter the clearinghouse of your hippocampus: a *trash bin* outside the door for toxic thoughts, a *revolving door* to quickly shunt any toxic thoughts that make it past the trash bin back outside, and the *memory bank* where you deposit only pleasant memories into your growing happy-brain account.

WHAT NEUROAUTOBIOGRAPHY DO YOU WANT IN YOUR BRAIN?

Whenever we dwell on a thought (happy or sad), a specific group of neurons related to that thought in our brain fire and wire together. That's how our experiences and thoughts become, in neurospeak, "encoded" or implanted in our brains, creating memories. Those memories are then stored in your brain's file cabinet. The more often a set of neurons fire and wire together, the more those neurons stay together, the more files you put in that memory's folder, and the thicker the folder gets. Those files become your neuroautobiography. What kind of thoughts do you want to store?

ENJOY NINE STRESSBUSTERS TO BALANCE YOUR BRAIN

Let's put think-changing your brain into practice by using these nine tools. Now that you understand how you can think-change your brain, for better or worse, and why it's so important to grow the happy center of your brain to reduce stress, we will next empower you with our top time-tested stressbusters.

PILLS SOMETIMES HELP YOUR SKILLS

"Doctor, can I just pop a pill to quickly feel better?" Yes and no! Most mood-mellowing medications take a few weeks to start working. Also, the mood-mellowing medicines you make yourself are safer and faster-acting than the medicines you take. Frankly, in our experience, pills without skills never work and never last. For these reasons, throughout this book you will learn a pills-and-skills model. First you try skills. Sometimes you try pills and skills, since pills may be necessary to jumpstart your program to ease you into the skills. But we recommend never using pills alone. (See more about the smart use of mood-mellowing medications, page 221.)

---

1. IF YOU CAN'T CHANGE IT, DON'T WORRY ABOUT IT

Stress researchers confirmed this "don't worry, be happy" teaching when they studied a group of centenarians (who not only live to be at least 100 years old, but live healthily and happily). One of the main "medicines" these centenarians reported was: "If I can't change it, I don't worry about it."

Dr. Vince once asked his live-in psychotherapist (a.k.a. his wife, Dr. Gayl Hartnell) what she believes is the number-one cause of the mental unwellness epidemic. Her wise reply: "We overthink about things over which we have no control."

Don't dwell on "dumb" decisions. Suppose you made an unwise decision: "I said no to this investment I should have made." Don't dwell on it. Beating yourself up by thinking "If only I had . . ." or "I should have . . ." not only robs your brain of energy it could be using to grow your happy center, it also raises your stress hormones. Often your brain gets more stressed out by your *reaction* to your perceived mistake than by the consequences of the decision itself. The more you dwell, the larger your worry center grows. The more you try to trash those negative thoughts, the smaller your worry center becomes. Take it from Dr. Bill.

DR. BILL'S HAPPY STORY

One day while on our favorite family vacation, boating in the Caribbean, we had mechanical problems on the boat we chartered, and we limped into the harbor for repairs. The cheerful boat repair person greeted us with "No problem, mon." That cheerful greeting rewired my brain toward: "Don't worry about the problem. Enjoy finding the solution." The rest of that week I wore my favorite t-shirt, which reads, "Don't worry, be happy." In fact, when we do our Caribbean boating adventures with guests, especially non-sailors, I send out a little pre-trip prescription that essentially says: "Only positive thinking, please. Something will always happen; we just don't know what. We focus on solutions and don't dwell on problems."

On one trip all of our guests were psychiatrists, psychologists, and neuroscientists. I felt like I was in continuous analysis. One sailing afternoon one of our guests became a bit negative, worrying about the weather forecast. Behind the sailboat we tow a small rowboat for our transit to shore, called a dinghy, that was riding along uncomfortably in the choppy water. Martha pointed toward it and quipped, "That little boat is our worry boat. The big boat is a happy boat. Anybody want to sit in the worry boat?" All the therapists smiled at her point.

2. TRASH TOXIC THOUGHTS

We've all heard about toxic chemicals affecting the lungs, the gut, and the skin, but fascinating neuroscientific research has uncovered another "toxin" that can affect the brain—toxic thoughts. Think of the prolonged high stress that dwelling on negative and toxic thoughts produces as spraying a toxic chemical on your brain garden, making the plants wither. Alternately, think of each of your thoughts as a tree. The more frequently and longer you think a thought, the deeper its roots become and the taller the tree grows. As you keep thinking about it, the tree grows branches, and the branches start connecting with neighboring trees, affecting how happily or unhappily these trees grow. You grow your own thought forest. Your thoughts control whether you get the beauty of a happy forest or the sadness of an unhappy forest.

The tree analogy is not a perfect one because, unlike a forest, which takes decades to grow and change, you can change your brain in a matter of months. What changes the quickest are the *connections* between the brain's thought trees: the growth of new connections and the pruning away of connections you stop using. In the lingo of neuroscientists, this is called "rewiring your brain." Some call it "renewing your mind." By quickly trashing toxic thoughts, you prevent them from taking root and growing.

People who have mastered this thought control quickly say to toxic thoughts, "Go away," "That's not me," "I won't do this," or *"Think this, not that."* The techniques in the next section can help.

COACH HAYDEN NOTES. . .

One of my most effective tools to trash a toxic-thought cycle when I just can't seem to do it on my own is to phone (or text) a friend. I have a few close friends and family that I can reach out to at any time and say "Talk to me about happy things," "Please pray for me about . . . ," or "Help me process this." Having these trusted voices guiding me back to healthy thought patterns is hugely valuable, and being that support in return leads to deep, intimate relationships.

Repel "ANTs." ANTs stands for *a*utomatic *n*egative *t*houghts, a type of toxic thought that suddenly and often repeatedly appears and catches you off guard. You don't always have control over the ANTs coming in, but you do have control over what you let stick around and what goes in the trash bin.

Master this "ANT repellent" technique: As soon as a negative thought invades your mind, brush it away like you would a pesky ant. Then immediately think of how to replace that ANT with an anti-ANT: switch to a positive thought to overshadow the negative one. Mentally *replay* a peaceful, pleasant visual memory from your "happy file" storage, such as your wedding day or a favorite vacation memory.

Remember the mood-switching scene from the movie *Happy Gilmore*? To rescue Happy from his negative, quick-to-anger mind-set, his girlfriend-turned-therapist smilingly advises, "Happy, go to your happy place." Happy then imagines his dream of being a star hockey player scoring the winning goal. You can repel ANTs the same way.

One evening, while Dr. Bill was watching a movie, a shot-soldier scene popped up. He reflexively closed his eyes, but the scary scene still got through. Next, to keep the toxic scene from sinking deeper, he mentally told himself, "Change channels," and replayed a happy scene: "Sisters, Sisters" from *White Christmas*, a song his and Martha's kids, sisters Hayden and Erin, sang at their fiftieth wedding anniversary party. ANT repelled.

While writing this section, we helped a friend repel the ANTs of his messy divorce. His joystick was his job. As soon as the divorce ANT appeared, he quickly repelled it by trashing the thought and replaying happy scenes from work.

Cognitive behavioral therapy. A common element of therapy, CBT simply teaches you to redirect your thinking—to quickly switch from negative to positive, and throw negative thoughts in the trash bin. We prefer to call CBT "think-changing your brain" (TCB), because by making a habit of changing the way you think, CBT actually changes the brain circuits, rewiring them in a positive direction. It puts different pathways in the brain, such as bigger pathways to the happy center and to the trash bin. This is why it's more likely to be permanent.

CBT works by questioning ANTs such as "He probably doesn't like me," and replacing them with more truthful thoughts, such as "I don't know if he likes me or not." Then you start thinking deeper and healthier: "Does he know me well enough to make that judgment? Do I want to give this person the power to make me feel badly about myself? No!"

In stressful situations, CBT teaches you to cool it for a moment, take a few deep breaths, and turn on your "stressless/rational thinking" system to better organize your thoughts and feelings and select a well-thought-out response that promotes an overall feeling of well-being. Then CBT shows you how to follow this reaction with nurturing self-talk and self-compassion ("I'm proud of myself. I didn't lose my cool this time. I'm getting better at this . . .") that leads to better self-esteem and more self-confidence. Over time, you will have think-changed your brain to replace ANTs with *automatic positive thoughts* and a stressless brain.

Put your mind on "mute." To practice repelling ANTs, try our favorite mind game, "mind-mute." Whenever your mind is "racing" or "scattered" with too much traffic, retreat into a quiet place and think "mind-mute"—your cue to declutter and clear your mind of negative thoughts. Partnering this mental clearing with *deep breathing* (see technique on page 141) helps you refocus more quickly. Initially, it may feel like you are "working" too hard to clear your mind, which could rev up your mind even more. Eventually, however, your mind will welcome the "mind-mute" cue and dutifully dial down disturbing thoughts.

To master mind-muting:

- Go to your favorite peaceful place where you are least distracted, such as a swimming pool, the golf course, or in the woods. If you can't go there, find a quiet space and visualize that peaceful place.
- Take a deep breath.
- Do a *five-second* mind-mute. This is usually the longest brain-cluttered beginners can go before disturbing "stuff" (thoughts) reinfects your peaceful rest.

- Gradually increase your mind-mute duration to ten seconds, then fifteen, and so on, until mind-mute naturally becomes easier.
- Practice mind-muting several times throughout the day.

CRY OUT YOUR STRESS

Tears have been found to contain the stress hormone cortisol, as if tears are the body's overflow for excess stress. Perhaps stress-releasing tears give more meaning to the phrase "a good cry."

---

3. TAKE A DEEP CLEANSING BREATH

Deep breathing increases the pleasure hormone dopamine and calms your mind by decreasing blood flow to the worry centers of the brain while increasing blood flow to the relaxation centers.

At least five times an hour, especially when you are feeling hurt or stressed:

1. Breathe in deeply from the belly, through your nose. First, feel your belly expand, then your chest. Exciting research reveals that nasal breathing triggers the lining of the nasal and sinus tissues to release nitric oxide (NO), the same natural healing and anti-inflammatory biochemical that is stimulated in blood vessels by exercise (see page 99).
2. Hold your breath for a count of five.
3. Exhale slowly for a count of five to seven, either through your nose or through pursed lips. Both keep the lungs expanded longer to deliver more healing oxygen to your brain. While exhaling, to help empty your lungs (which then allows you to inhale even more air), suck in your navel toward your spine.

Deep breathing variations. As you exhale, try *humming* with your mouth closed, which causes the air in your nasal and sinus cavities to oscillate more and vibrates the cilia of the nose to release more healing, relaxing NO. Words that end in an "m" sound are natural for humming with your lips closed. Try "Love (inhale) Mommmm (exhale)." Or "Love (inhale) himmmm (exhale)."

To enjoy another "NO effect," add deep yawning to deep breathing. Yawn with your mouth wide open like a big letter "O" during your deep belly breath. As you exhale, make a continuous sighing sound, such as "ahhhhh."

Dr. Bill likes to add a two-word mantra: He thinks "feeeeel" on the inhale and "goooood" while humming the exhale.

BREATHE BETTER TO EAT BETTER

Belly breathing isn't just good for your head brain; it's good for the gut brain, too! Do some deep breathing before and several times during meals. By stimulating your PNS, deep belly breathing facilitates digestion and dials down overactive stress hormones, which are notorious for compelling us to both make unhealthy food choices and overeat.

---

4. HAVE AN ATTITUDE OF GRATITUDE

Gratitude is about being thankful and appreciating what you have and receive from others. One of the most powerful ways to get out of a "stress funk" is to focus your thoughts on the gifts that you have at this very moment. Dwelling on those not only grows your happy center, but also shrinks your sad center.

No matter how difficult life seems, we all have a few blessings to be thankful for. Start simple, with something like: "I have a roof over my head, food and water, and people who care about me." Slowly, you'll shift from self-pity to serenity.

THE GRATEFUL BRAIN

Brain-health specialists have long recognized the link between how grateful a person is and their level of body and brain health. Neuroscientific discoveries now tell us why. People with the most grateful brains tend to be:

- less anxious and depressed
- generally more joyful (neuroimaging studies of the brains of habitually grateful people showed that the happy centers of their brains lit up more brightly)
- more successful, grade-wise, in college
- happier in social relationships
- better at stress management
- healthier, with lower blood pressure, heart rate, and levels of inflammatory blood markers (sticky stuff)
- better sleepers

So, if your therapist inquires, "Tell me three things for which you are the most grateful," you will appreciate the path they are taking you down. And if you are suffering from an illness, such as arthritis or heart disease, you may benefit from planting in your brain how grateful you are for those organs that are working well: "I am thankful for my eyes that see, my legs that walk," and so on. Being grateful for the things that are still working and focusing less on the parts that don't is especially helpful as we age!

Practice "I am . . ." meditations. A mother in the Sears' medical practice uses this tool to help her ten- and twelve-year-olds calm their sensitive minds at bedtime: "I taught them to quiet their minds as they drift off to sleep by thinking or saying, 'I am pretty . . . I am smart . . . I am a good soccer player . . .'" The younger you can learn self-regulating and self-calming tools, the better—this can turn into a wonderful lifelong habit.

See additional gratitude tips: flotation therapy, page 154, and "Think It," page 180.

FIVE THINGS I LIKE ABOUT ME

In his medical practice, Dr. Bill suggests that down patients paste on their bathroom mirror and cell phone wallpaper a list of their top five physical, emotional, and social assets. For example:

1. I like my job.
2. I have pretty hair.
3. I can swim fast.
4. I have two wonderful children.
5. I get up in the morning and nothing hurts.

Quickly replaying what you have dials down your stress about what you don't have. Remind yourself throughout the day as needed!

---

5. MANAGE ANGER BEFORE IT MANAGES YOU

Because anger is one of the top stressors that shrinks your brain, here's a crash course on how to manage anger before it manages you.

What Happens In Your Brain When You Get Angry

Your emotional center, which includes the amygdala, has a reciprocal relationship with your frontal lobe, the logical and rational center. When you get overemotional, whether angry, fearful, or hyper-anxious, the amygdala dials up and the frontal lobe dials down. Anger takes over, empathy and compassion decline, and you lose the ability to accurately assess how other people feel. As a result, while angry, there's a greater chance you'll say things you'll regret; you're less likely to make rational decisions, and more likely to make regrettable ones.

Anger disrupts your brain's ability to compassionately evaluate and respond in social situations—and left unchecked, it can do so permanently. In the moment, anger dials down the empathy center, which prompts you not only to act irrationally and make harmful decisions, but also to lose the awareness that you are being a jerk. You become self-centered, unable to listen and sense others' fears and needs. And over time, the neurochemicals released by persistent anger gradually "shrink" the empathy center of the brain, impairing your ability to think and act with compassion even when you aren't actively angry. Your anger becomes "you," sadly giving rise to the label, "He's such an angry person."

Fortunately, you can stop anger's cascade of brain-draining neurochemicals by quickly switching gears and letting your rational lobe put the brakes on the anger area. By saying to yourself, "Cool it! Think through what you're about to do," taking a deep breath, and rethinking the situation, you can balance your emotional center so that you're better able to respond rationally and with empathy.

"INFLAMED WITH ANGER"

Many people, especially those already suffering from immune system dysfunction, experience a flare-up of their pains, such as arthritis, following an angry outburst. Uncontrolled, prolonged anger floods your brain with biochemicals that lead to *neuroinflammation*, the root cause of most brain illnesses. Following anger with forgiveness may quell these flare-ups.

Calm, cool, and collected. This classic wisdom summarizes anger management. First, find your calm. Count to five and take a deep breath to dial down your anger and put your frontal lobe back in control. Second, stay cool. Finally, get collected. Take the time to collect your thoughts and empathetically imagine how what you want to say and how you say it may make both you and the other person feel.

One day, Dr. Bill asked Hayden's fifteen-year-old daughter about what she had learned while reading this chapter. We love what she said:

Don't speak when you're mad,
And don't decide when you're sad.

COACH HAYDEN NOTES . . .

Don't text while angry! So much of our communication these days is through texting and other messaging. Engaging in angry text conversations is not only a counterproductive communication method, it can also easily trigger us to write something we would never have said in person. When I feel myself getting angry during a text conversation, I either put my phone down entirely or switch over to a notes app, where I can get my angry thought out without sending it. Then I can decide later how to edit my response to best turn the conversation around.

Our Time-Tested Tips For Better Anger Management

Here are our top recommendations for anger management:

Tame the trigger. Identify what tends to set you off or light your fuse. For us, hunger and tiredness play a big role, so we try not to go hungry (lest we get "hangry") or enter into potentially anger-triggering discussions before bedtime. Social media and internet overload can upset your calm and trigger anger. Know when to turn it off. Alcohol, while a relaxant for some, may trigger anger for others.

Interrupt the flow. Don't be a knee-jerk responder. This means preprograming your moody mind to take time between experiencing an anger trigger and reacting to it. Prolonging your reaction time gives your rage center a chance to dial down, better enabling you to offer a rational response. Afterward, continue your programming by turning on your happy center and reinforcing your good choice: "I really handled that well . . ." or "I mastered my emotions . . ."

Walk it off. If you're struggling to delay your reaction time, try quickly walking away from the scene of the crime-to-be, your anger outburst. Moving your body helps move your mind away from your anger response. Walk outside, look up and say "hi to the sky," smell the roses. Just remember to tell your loved ones what you are doing first so they understand. Several times, Dr. Bill has said to Martha, "I'm going outside for a walk before I shout something I will later regret."

Shield your anger center. Quickly think "no access allowed" to angry thoughts. Imagine wrapping a protective shield around your anger center, the amygdala, to prevent it from firing brain-draining neurochemical bullets (stress hormones) all over your brain. This will help keep you from thinking and acting irrationally. (See related tip, "mind-mute," page 140.)

While writing this section, Dr. Bill suffered a distressing phone call from a jerk. Midway through the call he felt himself getting angry, triggered by the caller unloading his anger onto him. So, he put a brain barrier around his anger center. He listened, but didn't let his brain be bothered. He put the brakes on his anger center while accelerating his rational thinking and empathy centers. After a few minutes, he felt his mind shifting to focusing on the caller's need for help rather than on his own anger. At the call's end, instead of feeling the "anger effect," he felt a big dose of the helper's high.

Remember how persistent anger gradually "shrinks" the empathy center of the brain? Fortunately, the reverse is also true: the more you think about or exercise compassion, the more you increase the blood flow to and neurochemical stimulation of the empathy centers in your brain, making you less prone over time to outbursts of anger. One of the best ways to do this is meditation.

6. MEDITATE MORE, AGITATE LESS

The happiest people often are the most mindful meditators. Remember our stress-reduction theme: grow your happy center and shrink your sad center. Meditation does both.

Contrary to popular belief, the brain doesn't really shut off when it meditates. Rather, meditation quiets the mind and blocks out mental clutter, the business that contributes to being a "scatterbrain." Brain scans reveal that during meditation, the busy center becomes less busy and the emotional center becomes more calm. Here's another way of looking at meditation, as described by Dr. Richard Davidson, a pioneer in brain imaging during meditation: "The brain on meditation is less 'bothered' by the usual stressors and is more accepting and content."

Perhaps the biggest meditation misconception is that it is mainly for monks. On the contrary, this ancient art of mind healing is just what the brain doctor ordered for today's supercharged, plugged-in minds.

Why Meditate?

Remember our mantra for brain health: *better (blood) flow, better brain.* Meditation increases blood flow to the areas of your brain devoted to relaxation and calmness.

We now live in a world of extreme multitasking, which overloads our minds and drains our brains. Meditation, an age-old remedy for this modern ailment, enables us to unload, at least temporarily, some of the excess baggage we carry. Meditation is your mind prompting, "Enough already, I need a break!" Learning to enjoy more daily meditation therefore makes sense.

YOUNG ADULTS NEED MEDITATION, TOO

While meditation is good for all ages, we believe twenty- and thirtysomething adults can benefit the most, because the entertainment, advertising, and social media industries are all competing most fiercely for younger eyeballs. This group needs the space to make clear decisions about still-developing careers, relationships, families, and personal growth. Meditation can provide this space by helping you more selectively process all the constant "pings" at work, on your phone, and while trying to relax at home.

Meditation makes you more compassionate. As you learned, meditation increases blood flow to the compassion center of your brain. Neuroimaging recordings reveal that during what is called "compassion meditation"— where the meditator focuses on compassionate thoughts for another person—the brain's compassion/happiness center (the hypothalamus) lights up. What's even more interesting is that, even after the meditation finishes, the same center stays lit longer and more intensely, making you more prone to compassionate action and better able to manage stress.

Meditation makes happy hormones. During deep and habitual meditation, *serotonin* levels may go up, which is why meditation is a skill that is equally effective as or even more effective than some pills, while being safer. That makes sense, since meditation may help put the upset brain back into biochemical balance through a similar mechanism as medication, but in a more personalized way. Meditation is medicine for *your* mind. Meditation may also trigger higher levels of *dopamine*, which helps explain the sense of relaxation, happiness, and peacefulness that meditators experience. In addition, meditation lowers stress hormone levels. This drop in stress hormones and rise in calming hormones may also explain why meditators are better able to navigate healthfully through stressful situations and often make better decisions.

Meditators enjoy good gut health. See chapter three to learn about the relationship between stress and gut feelings, page 87.

Meditators live in the moment. The cluttered mind often creates stress by dwelling on the unpleasant past or trying to predict the uncertain future. Meditation trains your mind to focus on the here and now. During meditation, the meditator focuses on the present environment without judgment or executive evaluation. Freed of worry about the past and future, they are better able to savor the present moment. PET scans reveal why this happens: During meditation, blood flow decreases to the judgment and worry centers of the brain, while increasing to the sensation and enjoyment centers.

MEDITATION: MEDICINE FOR THE AGING BRAIN

Long-term meditators show less age-related "cortical thinning," meaning they lose less gray matter and their brains shrink less with age.

While you can't always control the thoughts that enter your mind, you can keep them from overtaking it. That's what meditation does. It mutes negative self-talk and concerns about the past and future, sort of like purging your inbox of junk mail.

Sound like what you need? Read on.

How To Meditate

The best way for you to meditate is one you will consistently do. Try these time-tested tips on your way to formulating your *personal peace plan*.

COACH HAYDEN NOTES . . .

There are many fantastic free and subscription-based apps and websites that can help you start and maintain a meditation practice. These can make meditation less intimidating and more accessible, and can expose you to a wide variety of meditation practices. My favorites are the guided meditations where a soothing voice prompts my breathing and thoughts. I can choose from numerous categories, such as stress relief, inner peace, focus, anxiety, sleep, and gratitude, with time frames ranging from two minutes to an hour. I can even set reminders on my phone to stop and take a "five-minute mind break" during the most stressful parts of my day.

Meditate at dawn and dusk. For most people, the two best times to meditate are shortly after awakening and just before going to bed. Enjoy those two "happy hours" whenever you can.

Meditate to derail stress. When you feel yourself ("self" meaning mind, brain, and body) getting stressed, immediately tap into your brain's "happy file." It contains your happiest moments and memories, and will quickly reroute your thoughts from worrying to dwelling on a happier path.

Practice meditation poetry. Dr. Bill's favorite mini-meditation is to replay a poem from his memory file and liven it up by replaying the scene that sparked the poem. One day as he and Martha were strolling in the woods and feeling the relaxing effects of the trees in nature, he looked admiringly at Martha and said:

I feel that I shall never see,
a lady lovely as you are to me.

After replaying the poem, he feels unstressed.

Develop a personal mantra. A mantra is a word or phrase that protects your mind from sabotaging thoughts. During meditation, your mantra helps you quiet the clutter and the chatter of self-talk in your mind by giving you something simple to focus on. Once you've developed your personal toolbox of favorite mantras, as soon as a stressor arises, clicking into one of your mantras is like sending a text to the relax-and-cool-it center in your brain.

One of Dr. Bill's favorite mantras is a simple prayer, "Guide me, God," repeated frequently throughout the day as a mini-prayer during a mini-meditation. It helps quiet his mind during decision-making challenges.

Enjoy musical meditation. We enjoy "M&Ms" for health: not candy, but "music and meditation." Using music in meditation enables you to mellow the mind by going with the flow and rhythms. Music can be both a mental calmer and exciter, so mind-match the music you need for the mood you're in.

While the best music to use is a personal choice, some of Dr. Bill's favorite musical pieces for meditation are:

- "The Prayer," by Celine Dion and Andrea Bocelli
- Emile Pandolfi piano medleys
- "Moonlight Sonata," by Beethoven
- "What a Wonderful World," by Louis Armstrong
- "Somewhere Over the Rainbow," by Israel Kamakawiwoʻole
- "Hawaiian Wedding Song," by Elvis Presley

YOUR BRAIN ON MUSIC

One of the newest interests in brain health is called the *neuroscience of music*, which explains why music is so therapeutic for the brain. Listening to brain friendly music triggers relaxing brain waves similar to those you get as you begin falling asleep or during relaxing meditation. Music also pleasantly diverts our attention away from worrisome thoughts, toward rhythmic sounds that register "like!" in our auditory center.

Enjoy moving meditation. Your brain loves the feel-good effect of increased blood flow, so synchronizing your mantra with a rhythm of movement, such as walking or dancing, is a great way to boost the benefits of meditation, especially for meditators like Dr. Bill who have a hard time sitting still. Better still is movement in nature.

Need a movement that's easy on your body and benefits your brain? Swimming is a magnificent movement for your mind. It can be more calming than walking because you can relax your whole body.

You can also walk *in* water. In a peaceful pool, immerse yourself shoulder high and enjoy the floaty feelings of a slow, rhythmic, meditative walk. Time the touch of each step with a calming word: "Thank (step) God (step) I'm (step) healthy (step)." Or: "Walk (step) in (step) wa (step) ter (step)." Synchronizing your mantra with your natural stepping rhythm calms your mind.

DR. BILL'S TOP MORNING STRESSBUSTERS

I begin most mornings with a "meditation swim" or "prayer swim" in an outdoor pool. While swimming, I put my mind on "dial down" mode, trying to avoid all clutter in the brain, and simply let the sights and sounds of nature sink in. While doing the breaststroke, I look at the blue water under the green trees and the blue sky while praying, "Praise God," during one stroke, and then adding, "I'm happy," during the other stroke. The rhythm of the body and mind moving and praying together is very calming. (Watch a video of Dr. Bill demonstrating this attitude of gratitude meditation at AskDrSears.com/meditation.)

Treading water to shed worries is another one of Dr. Bill's favorite relaxation techniques. Dangle your body in the deep end of a pool. Move your arms just enough to keep yourself afloat while your legs dangle weightlessly, flutter-kick slightly, or just tip-toe touch the bottom of the pool. Close your eyes and either "mind-mute" (page 140) or meditate on your mantra. As your body enters this almost weightless state, you will notice that your mind also sheds the "weight" of worries.

7. BELIEVE IN A "HIGHER POWER"

That's why I say, with the utmost confidence,
that God can change your brain.
–Andrew Newberg, MD, neuroscientist and coauthor
of *How God Changes Your Brain*

Can you *believe* your way to better brain health? Yes! Over our fifty years as doctors, husbands, fathers, and grandfathers, we have noticed that the deeper a person's spiritual health, the happier their mental health.

"How's your spiritual life?" may now be a leading opener in therapy sessions, as more and more brain researchers and healthcare providers are realizing spirituality's therapeutic importance. Ruminating on who, what, and where God is continues to challenge our mere human mind. Yet thanks to greater scientific understanding of the human brain, neuroscientists are now able to measure what we call *the God effect,* and some call "the belief effect": how the more deeply we feel connected to God and directed by God, the happier and healthier we tend to be. This new science, called *neurotheology,* shows that our spiritual health can greatly heal our mental and physical health. Neuroscientists have proven that people who grow a larger God center are more likely to:

- enjoy more mental wellness.
- enjoy more physical wellness.
- enjoy faster healing from illnesses, injuries, and accidents.
- experience less depression.
- remain steadier during anxiety episodes.
- not only live longer, but stay sharper longer.

Who wouldn't want that?

The "wow!" factor of why believers enjoy healthier brains is the discovery through modern neuroimaging techniques that the areas in the brain that help dial down stress—mellow moods, calm worries, and direct happier thinking—are larger and smarter in believers. Spiritual beliefs and practices increase your empathy or compassion center, the anterior cingulate, and quiet an overactive anxiety center in your amygdala. Yes, the God effect is real and can structurally change your brain for the better.

YOU CAN REGROW CHILDHOOD SPIRITUAL ROOTS: DR. BILL'S SPIRITUAL JOURNEY

I grew up in an emotionally rich but financially poor family. Thankfully, my single mom guided me to plant deep spiritual roots at the early childhood stage, when my brain garden was most fertile.

Unfortunately, those roots later shriveled. During my medical training one of my most trusted and influential professors "mentored" me with this harmful advice: "God is only for those who need a crutch." My God center shrunk, as did much of my joy in life. When I let God leave my daily life, part of my brain garden withered.

Looking back at Martha and our fifty-three years of marriage, I realized that our marital life paralleled our spiritual life. During my spiritual withering I was oblivious to the fact that my letting go of God in our family also led Martha into God-withdrawal. I was failing in my role as spiritual leader of my family. When my God center withered, so did our marriage, which nearly ended in divorce until my self-centered living let God back in to water my withering brain garden. Soon after reawakening my spiritual roots, our marriage flourished. We added six more kids to our rediscovered spiritual garden.

The lesson I learned: Because spiritual roots were planted early and deeply in my life, they were more easily regrown later, even after years of letting them wither.

How The Belief Effect Benefits Your Brain

All three of us have noticed that our brains are happier when our God effect is stronger. When we sincerely feel and believe we are connected to God and directed by God, we are happier.

Wisdom from spiritual people we have interviewed. When someone shares how God helped them heal, or we see it in our medical practices—especially in dealing with circumstances beyond any earthly explanation—here's what we've heard that gave them health and hope:

"God pulls me up when I'm down."

"My faith in God lifts me up when I'm about ready to hit rock bottom."

"God gives purpose to my life."

"During my daughter's healing, there was something I could do as part of the medical team. I could pray to help her heal. That helped me to feel more in control of the situation."

DR. VINCE'S GOD-HEALING STORY

At the hospital where I was chief of neurology, one day a "code blue" summoned me to the emergency department. My four-year-old daughter, Kaylee, was being resuscitated after she stopped breathing during a continued seizure, called *status epilepticus*. She wasn't responding to the resuscitation, prompting the doctors to ask me, "Do you want us to continue?"— doctor-speak for, "We've tried everything and it seems time to let her go." I prayed as deeply as I could while insisting, "Yes, we will continue the resuscitation until I drop."

Thank you, God, for saving my daughter's life and healing her brain. She is now Dr. Kaylee.

Our world needs God. Growing your God center can help you be resilient when facing a world of terrorism, school shootings, and other acts of non-empathy. With the incidence of stress-related illnesses reaching their highest levels ever, it's time to let God in.

The more you contemplate God, the more you direct your thoughts and energies toward the role God plays in your life and what God means to you, the more you get outside of yourself and into another "person"— in this case, God. The more you get outside of yourself, the more you grow your empathy center. The process of wiring your brain for God can be summarized with one of the simplest sayings about the God effect: "Let go and let God."

You also realize your mere human frailties. By acknowledging God's presence in and guidance over your life, you don't feel alone, you don't feel out of control (you control your thoughts about God), and you don't feel helpless and hopeless.

How Growing Your God Center Balances Stress

While the actual "God centers" in the brain may be many and different for each individual, brain scans suggest the hypothalamus, your brain's air-traffic controller, is an important spiritual center. Meditators who are also more spiritually minded, brain scans reveal, have a thalamus that redirects worrisome thoughts away from landing all over the brain, enabling you to "don't worry, be happy!"

Another area of the brain that grows is the main "happy and social center" called the *anterior cingulate* (AC). This is the area that promotes empathy and compassion, that makes us "feel" for other people. Let's call the anterior cingulate the "feelings center" of your brain. It is strategically situated near the junction of your executive center (the "seat of reason") in the frontal lobe, where the brain creates and thinks, and the emotional center. Imagine the AC as another orchestra conductor of your brain's hormonal symphony that is directing you to "think smart, act wisely, and feel compassionately." Spiritual practices grow a person's compassion center in the AC.

Spiritual practices also quiet an overactive anxiety area in the *amygdala*. Wow! What a stressbuster!

Believe better, worry less, become more kind and calm. We all want that!

THE FOUR TOP MENTAL AND PHYSICAL EFFECTS OF GROWING YOUR GOD CENTER

During our years as doctors and observers of the mental and physical effects of spirituality, as well as reading what science says, here are the belief effects you, too, can experience by growing your God center.

Grow your God center, be healthier. Studies done by neurologists, gastroenterologists, cardiologists, immunologists, and just about every other -ologist all agree: the deeper your belief and practice of spirituality, the less likely an illness will occur.

Grow your God center, heal faster. Many doctors and scientists, and your authors, have witnessed beyond a doubt that patients with a more active God center heal faster from illnesses and accidents. We learned on page 131 about the *belief effect*—that the more deeply you believe you will heal, the more likely you are to heal. A personal belief that God will help you heal, as well as belief in the power of your social network's prayers, only intensifies that effect. Like food, prayer is good medicine.

Grow your God center, be happier. Neuroimaging studies have shown that not only do happy centers tend to light up in the brains of habitually spiritual people who pray and meditate often, but those centers are actually larger than in nonmeditators. Spiritual people also tend to have higher levels of happy hormones, especially dopamine, the neurochemical most associated with peace of mind.

Grow your God center, delay Alzheimer's. We doctors have long noticed that the more deeply spiritual a person is, the lower their chance of getting Alzheimer's, and the greater their chance of delaying Alzheimer's and lessening its severity. For many people, as they age, their brain tissue literally shrinks. A brain scan study done by the Psychiatric Neuroimaging Research Program at Massachusetts General Hospital found that, in some deeply spiritual people, not only was the usual brain shrinkage that occurs with aging delayed, but in some cases the gray matter actually thickened. Growing your God center helps preserve your gray matter as you turn gray.

Despite the greatness of our brains, humans may never fully know who God is, but we can experience a glimpse through the healthful effects we experience from our spiritual lives. That thought empowers us to grow our God center.

As we've seen, brain tissue, like muscle, follows the "use it or lose it" principle. The more you exercise muscle, the larger and stronger it grows. Thoughts and the energy that they consume are to brain tissue what

exercise is to muscle. The more you think, meditate, ponder, or whatever you want to call it about God, the larger and deeper your God center
grows. It's as spiritually simple as that.

8. START YOUR DAY THE PEACEFUL WAY

How you begin your day often sets the tone for how you spend your day.
Recall from chapter two how people who start the day with a healthier
breakfast tend to eat healthier the rest of the day, a smart practice called
frontloading. This concept can also help you enjoy healthier thinking. You
can frontload, or what we also call *preload*, your brain by starting your
day with your favorite and most peaceful thoughts.

The five-minute way to start your day. As you ease out of bed, immediately
click into an "attitude of gratitude" mind-set, with a thought such as: "I
am thankful for . . ." or "Today I look forward to . . ." Pick five things you
are thankful for and start the day focused on them. Remember, the more
you think about a subject, the more blood flow to that center of the brain.
Starting the day this way is like telling your brain, "Hey, gratitude center,
light up and start the day a happy way." At least for the first five minutes
of your day, trash any toxic thoughts or worries about undesirable tasks
you have to do but don't want to, in favor of a brief meditation on the
things for which you are most thankful.

Let the sunshine in. To start your day in a brighter way, immediately go
from your bed to a window and look or go outside. Light up your mind by
filling it with bright thoughts and scenes, leaving little room for "dark"
thoughts. Next, spend at least five minutes meditating, either quietly or
in movement meditation.

Brighten another's day. After you've done your "five-minute way to start
the day," think: "What can I say to the first person I meet to brighten
their day?" Maybe that's a member of your family. Dr. Bill and Martha
greeted their kids, and now their sleepover grandkids, with "Good morning, sunshine!" Or the first person you meet may be a server in a coffee shop, a stranger on the subway, or someone on the street who looks

troubled. A smile or kind word might be all that person needs to stop dwelling on their troubles and think about what's good in their life. (See related section, "Dr. Bill's Top Morning Stressbusters," page 153.)

9. ENJOY THE HELPER'S HIGH

Feeling good while doing good is what we call the "helper's high." When you help someone get happier or healthier, you turn up your feel-good neurohormones. One of the best ways to relieve your own stress is to help others relieve their stress.

Use the nine tools from this chapter to formulate your *personal peace recipe*, hopefully to last from sunrise to bedtime. As you use our healthy brain plan to invest in your mental health savings account, if someone asks you what change you have made that you found the most challenging, yet the most healthful, you may find yourself replying, "I mastered stress management!"

NOTE TO THE READER

This chapter naturally feeds and flows into chapter seven, "How to Heal from Anxiety and Depression." You may find that reading these two chapters together gives you even more self-help tools to manage your stress.

CHAPTER 6

$\sim\sim\sim$

SLEEP SMART: HOW TO GET A GOOD NIGHT'S SLEEP

"Good night!" Remember hearing those sweet words from your mom or dad as you drifted off to sleep?

Like food, sleep is medicine. The main goal of our book is to give you the tools you need to keep your brain in balance. That's what a good night's sleep does. During a good night's sleep, our brains get relief from all the "stuff" that bothers them during the day. Also, your brain and body are "detoxing" the toxic thoughts and chemicals that infected you during the day. By dialing down stress neurochemicals and dialing up calming ones, good sleepers enjoy a balanced emotional brain. Wakeful sleepers don't.

In this chapter, we will empower you with tips to develop your own peaceful sleep recipe to open your personal brain-health pharmacy while you sleep. You spend 30 percent of your life sleeping. Enjoy it!

OPENING YOUR PERSONAL SLEEP-WELL PHARMACY

Think of sleep as your personal nighttime pharmacy, open from dusk until dawn. Here's why sleep is good medicine for your mind:

Daytime Problem	Nighttime Sleep Solution
Stress	Stress dials down
Brain-garbage overload	Brain unloads "garbage"
Mental fatigue	Brain reenergizes
Worry	Don't worry, just sleep!
Hurts	Hurts heal
Eyestrain	Eyes rest
Problem overload	Problems are solved
Worn out	Brain and body refreshes
Depressed	Natural antidepressants released

Sleep better, heal better!

SLEEP AWAY DEPRESSION AND ANXIETY

Sleep scientists call stress the "anti-sleep virus." Think of your bedroom—your sleep sanctuary—as your "therapist's office." One of the first questions we always ask our patients is: "How's your sleep?"

Good sleep, good emotional balance. Smart sleep is your personal in-home therapist for emotional wellness.

Doctors, and parents, have long noticed that a poor night's sleep leads to emotional instability the following day. We're more likely to "fly off the handle," become angry, and overreact to life's little setbacks.

A good night's sleep balances the brain's emotional accelerator (the amygdala) and its emotional brakes (the frontal lobe). Brain scans comparing a person's emotional response following a good night's sleep or a bad night's sleep showed that the *amygdala* is greatly amped up in people who are sleep deprived. This amped-up amygdala wears out the thinking center, the frontal lobe, keeping you from waking up "refreshed" and able to think clearly. A worn-out frontal lobe means your brain has weaker brakes, which allows the amygdala to accelerate out of control. Yes, following a poor night's sleep, you may wake up literally "out of control." A good night's sleep helps you better control your emotional reactions the next day.

Good sleep leads to good moods. Depression is caused by hormonal imbalance, and much of this "imbalance" occurs during poor-quality sleep—making good sleep your safest "antidepressant." When you're asleep, the brain releases more serotonin, the happy hormone, along with dopamine, the pleasure hormone. When you don't get enough sleep, or when the sleep you get is poor quality and interrupted, your brain isn't able to make these natural antidepressants. *Quality sleep* means your sleeping brain is in biochemical balance—that it's received the best proportions of deep and light sleep stages (see page 170) and a sufficient number of hours to feel refreshed.

Quality sleep does two other things that lead to better moods. It prevents your SNS, your nervous system's rapid-response team, from overreacting to stress. (See SNS/PNS balance, page 122.) It also triggers melatonin, which we call the "happy sleep" hormone. Better hormonal balance during sleep leads to happier moods the next day.

Sleep better, forget less. Some of the medicines your brain makes while asleep are *memory medicines*, which is why forgetfulness is one of the most noticeable problems from lack of sleep.

Picture your brain like a busy office, in which a lot of information files were made during the day. During sleep, your brain—your giant walk-in file cabinet—sorts your daytime information, tweaks and prunes it a bit, and then files it in the memory files in which it belongs: names, phone numbers, information you study for a test, and so on.

Most of your daytime information is initially stored in your short-term memory bank, your hippocampus. But this storage file is small, and can be easily overloaded. Every night during sleep, your brain shifts this information into smart file cabinets. When you awaken, the hippocampus has been partially unloaded of the previous day's information and memories, and is ready to take on new ones. Without a good night's sleep, your file cabinet may not have enough room to store newly acquired information—which is what leads to poor memory.

SLEEP AWAY TOXIC THOUGHTS

Some events, memories, and information you want to remember, and others you want to forget. Sleep facilitates both, by being wisely selective about what thoughts get stored and what thoughts get trashed. Because, as you learned on page 138, trashing toxic thoughts is at the top of the list for think-changing your brain, sleep is especially important for treating mood disorders. Yet mood disorders can often trigger sleep disturbances, which then make the mood disorders worse. It's important to note that some prescription mood-mellowing medicines have insomnia as a side effect, and this may be another reason why they don't work for some people—and may make some people feel worse.

SLEEP WELL, DETOX, AND PREVENT ALZHEIMER'S

Now the biggest public health challenge of the twenty-first century, Alzheimer's disease (AD) has increased as the number of nightly hours we sleep has decreased. Any correlation? Neuroscientists believe yes.

Garbage in, garbage out. One of the newest insights on the cause of AD is that it may be a garbage disposal problem rather than, or in addition to, a garbage-buildup problem. Here's how: The *glymphatic system* is the sewage network of the brain. It works like the lymphatic system in the rest of your body, but it's composed of glia cells. During sleep is when your brain's sanitation system revs up to do a *nighttime cleanse*, literally washing away neurotoxins, such as aluminum and amyloid (sticky stuff). And this cleansing power is highest during deep sleep.

What is even more fascinating is that, during deep sleep, the glymphatic system widens. Picture a street getting wider at night to allow more garbage trucks to clean up the daily mess. Less deep sleep means less removal of waste products, more storage of waste products, and poorer brain health.

Our conclusion: Sleep-deprived living, especially as we age, is one of the prime triggers of Alzheimer's.

MORE HEALTH BENEFITS OF SLEEP

Here's just a partial list of all the other "medicinal" effects that you get from a good night's sleep, which also affect your brain health:

Sleep better, leak less. No, not the bladder—the blood–brain barrier (BBB). As you learned on page 14, your BBB protects your vulnerable brain tissue from toxins. The weaker your BBB, eventually the weaker your brain. An interesting article in *The Journal of Neuroscience* in October 2014 reported that, at least in mice, chronic sleep deprivation can cause the BBB to leak.

Enjoy *heart*felt sleep. When sleep puts the brakes on stress in the brain, it also puts the brakes on how stress affects the heart. Poor-quality sleep triggers the sympathetic nervous system (SNS) to release the stress hormone cortisol, which strengthens your heart's contractions to pump more blood and accelerates your heart rate. Stress hormones constrict your blood vessels and cause atherosclerosis (buildup of sticky stuff) due to hypertension. During quality sleep, the parasympathetic nervous system (PNS) slows the heart down so the heart can rest. Less pressure is needed so the blood vessels and your heart last longer.

Sleep better, spike sugar less . . . and prevent diabetes. The brain needs a *steady* supply of sugar, not a spiky one. As you learned on page 21, "avoid sugar spikes" is Brain Health 101. But the rest of the body benefits from fewer spikes, too, and that's what quality sleep delivers.

Poor sleepers tend to have *higher sugar spikes* during the night and can develop insulin resistance, in which cells close the doors to insulin, leaving excess sugar (glucose) in the bloodstream and causing hyperglycemia, or hyper-sticky stuff. This excess sugar then ends up stored in the belly.

Quality sleep also helps you burn just the right amount of carbs. Yet, poor quality sleep dials down carb burning, which is what increases your glucose level at night and eventually leads to more fat storage around the middle. The longer your glucose level remains raised, the more insulin you produce. Eventually insulin resistance occurs. During quality sleep, your need for glucose dials down, which then dials down your need for insulin.

Sleep well, stay lean. The better you snooze, the more you lose. When you sleep poorly, it throws the neurochemical text messaging between your head brain and gut brain off balance. The higher levels of the stress hormone cortisol caused by a poor night's sleep appear to cause gut upsets and imbalance in our gut bugs. As a result, you tend to be more hungry and less satisfied, so you eat more the next day.

The reason you get hunger pangs after a poor night's sleep is due to a hormonal imbalance between leptin (the eat-less hormone) and ghrelin (the eat-more, hunger hormone). Sleep deprivation is a double-whammy: It decreases leptin and increases ghrelin, so you're never really satisfied and you keep on eating.

Also, because you make more melatonin during quality sleep, and melatonin triggers leptin, the more quality sleep you get, the less you overeat, and the leaner you stay.

In addition, brain scans show that the judgment centers in the prefrontal cortex tend to, shall we say, stay asleep in those who have poor-quality sleep the night before. Thanks to poorer impulse control, many poor sleepers end up eating an extra 500 calories a day, which translates into around a pound of extra fat a week.

Sleep well, boost your immune system. You tend to sleep more when you're sick because deep sleep strengthens your immune system to help you heal. The better your sleep, the more infection-fighting white blood cells your body makes. And when you're deprived of sleep, your immune system response is weakened.

SLEEP CANCER AWAY

While cowriting this book, Dr. Bill was diagnosed with a chronic, but curable, cancer called chronic myelogenous leukemia. Having lots to live for, he devoured heavy doses of whatever information he could use to heal. What he found was a sleep-healing discovery that even as a doctor of more than fifty years he did not know: *the sounder you sleep, the more powerfully your immune system fights cancer.*

We all have inside us a cancer-fighting army called natural killer cells (NK cells for short). These cells mobilize under circumstances that can seriously harm the body, and cancer is the main terrorist that they've been trained to fight. NK cells love sleep. In addition to helping you sleep well, melatonin triggers the immune system to produce more NK cells. The deeper and longer your sleep, the greater the quantity and quality of NK cells you produce.

Sleep also aids the body's natural cancer-fighting abilities in two other ways. Cancer cells feed on high blood sugar, which decreases when you get quality sleep. And sleep's ability to decrease stress by balancing the body's automatic nervous functions keeps your immune system from getting worn out or confused, making sure it's able to do its job to the best of its ability.

Thanks to expert medical care and good self-care habits— including a good night's sleep—Dr. Bill is now cured. Though his initial blood test showed almost *60 percent of his white cells* were cancerous, a year later less than *0.06 percent* of his white blood cells tested positive for cancer. (Our healthy brain plan helps preload your immune system to prevent and heal from cancer.)

ENJOY "BEAUTY SLEEP"

If you need more convincing, here's one more sleep perk: Sleep can delay skin aging. Just as sleep refreshes the brain by removing toxins and growing new tissue, so it also helps refresh and repair skin, including strengthening elastic tissue to help prevent wrinkling.

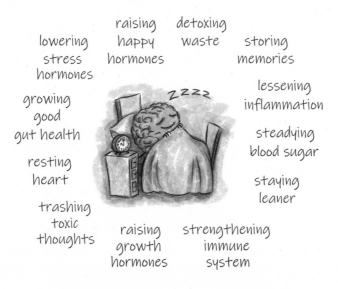

raising happy hormones

detoxing waste

lowering stress hormones

storing memories

growing good gut health

lessening inflammation

resting heart

steadying blood sugar

staying leaner

trashing toxic thoughts

raising growth hormones

strengthening immune system

Healthful Sleep at a Glance

ENJOY SMART SLEEP STAGES

Here's what happens inside your brain as you are "falling asleep." To help you remember the two main sleep stages, think the "two Ds": *deep sleep* and *dream sleep*. Each period of deep sleep, called NREM sleep, typically lasts ninety minutes. In between, we experience a short, light sleep stage of dream sleep, called REM sleep. This cycle of deep sleep and light sleep continues throughout the night.

When studying how to sleep more soundly, Dr. Bill initially wondered why we don't stay in deep sleep all night. Why do we have to "cycle" in between light and deep sleep? Turns out Mother Nature had a smart reason: each stage of sleep has its own night-shift brain-maintenance job.

The deep stage of sleep, which prevails during the first half of the night, is mainly designed to prune the brain of nerve connections made during the day that it doesn't want to keep. Dream sleep, which prevails the second half of the night, stores the connections that remain. Because the brain is so overloaded with information and problem-solving tasks during the day, it needs two stages to do its smart work: one to prune and one to build and store.

Let's take a closer look at the different "medicines" released inside your body and brain during these two sleep stages.

1. DEEP SLEEP

During deep sleep, your body and brain rest their hyped-up metabolism from the busy day and sort out what daily memories you want to keep.

Thank you, thalamus! The thalamus, as you learned on page 15, is the sensory gate of your brain. It decides which sensory signals are allowed through and which are not. The ones that gain access are diverted to the appropriate centers of the brain, which then make a conscious decision on what they mean and how to act on them. When you fall into a deep sleep, the thalamus shuts its gates, called a sensory blackout. Again, what a smart design, as if the conscious brain is saying, "I need a rest from all the sensory stuff you overloaded on me during the day." During sleep you are truly "in your own world," since your brain blocks out worldly interference.

Solve my problems, please. During deep sleep is when problem solving occurs. Suppose you have a problem, but your brain is so cluttered or scattered with daytime noise that you literally "can't think straight." In deep sleep, when your smart brain kicks out the clutter, it allows the pieces of the puzzle to fit together. You go to bed with a problem and you wake up with a solution. Thank you, deep sleep!

2. DREAM SLEEP

While its sleep-stage buddy, deep sleep, focuses on solving "intellectual" problems, dream sleep helps heal "emotional" problems. Sleep scientists call dream sleep "overnight therapy."

> REM sleep dreaming removes the emotional
> sharp edges of our daily lives.
> –Matthew Walker, author of *Why We Sleep*

How dreams heal. Dream sleep tells your anxiety center to "don't worry, be happy!" During dream sleep, our brain doesn't just store good memories from our waking hours. It also neutralizes bad ones. While the bad memories themselves aren't totally wiped away, what is wiped away are the stressful emotions that accompany those memories. The rational center of the brain puts the brakes on bad memories, smartly saying, "No problem. Yeah, that happened, but it's in the past, don't let it bother you." When we aren't able to accomplish this during sleep, it contributes to debilitating anxiety, which can keep you awake and resurface during the daytime hours.

Stress less while dreaming. REM dreaming stage 1 is the most de-stressing period of the day. During this stage of sleep is where your *stress hormones are lowest* and your brain is truly enjoying stressless sleep. That's why restless and stressless are at opposite ends of health.

COACH HAYDEN NOTES . . .

During our bedtime prayers, my children never forget to pray for good dreams. For them, anticipating sweet dreams is an important part of their nighttime routine, and leads to better sleep!

FOR SMART SLEEP, LISTEN TO YOUR SLEEP NAGS

For your nighttime storage system to work efficiently (storing the best connections and trashing the rest), your brain needs the right percentages of deep and dream sleep, which it can get only if you sleep during the period of time prompted by internal sleep-pressure neurochemicals such as melatonin and adenosine (see page 182). If you don't listen to your natural internal prompts on when to go to sleep—if, for example, you try to say, "I'll just stay up later to finish my work and then sleep in"—you seldom get the same quality sleep that you would have enjoyed had you listened to your "go to sleep" prompts and retired earlier.

So it's not just the total number of hours that you sleep that is important for good brain (and other!) health, it's *when those total number of hours jive with your internal sleep prompts.* This is why the classic sleep advice "You need seven to eight hours" is only partially correct. More accurate would be "You need at least seven to eight hours *during the hours that your brain prompts you to sleep and in the right percentages of deep and dream sleep.*" When these sleep prompts occur can vary among individuals. One person may get prompts to go to sleep at 10 PM and awaken at 6 AM; others may be prompted to sleep from 11 PM to 7 AM.

YOUR SLEEP-STAGE PHARMACY

Picture having a nighttime pharmacy with two aisles—a deep-sleep aisle and a light-sleep aisle—each one stocking different medicines that are dispensed during the night. Consider deep (NREM) sleep as making medicines for your *physical* health and dream (REM) sleep for your *mental* health, although each stage contributes to both.

Deep Sleep	Dream Sleep
• Lower heart rate	• Trashed toxic thoughts
• Lower blood pressure	• Lower stress hormones
• Better immune system	
• Stable blood sugar	
• Lower inflammation	

WELCOME TO SLEEP SYMPHONY HALL

You have a sleep symphony playing nightly in your brain. Mental wellness is the music of a well-balanced cerebral symphony. Your cerebral symphony plays twenty-four hours a day, but it has a different orchestra during the daytime than at night. During the day you have a few master conductors, such as insulin, adrenaline, and cortisol, that supply your muscles and brain with more glucose fuel. At night, a new orchestra appears for the night shift and the conductors are different: namely, growth hormone, prolactin, adenosine, and melatonin.

One of the daytime conductors that should start dialing down for the nightshift is *cortisol*. But when cortisol remains elevated, it continues its daytime duties: it revs up stress, triggers hunger, stores fat, and raises blood sugar—exactly the opposite of the "music" you want to be playing at night. And the poor-quality sleep that results causes your serotonin (the calming hormone) to decrease in your blood and, therefore, play less "comfort music."

Melatonin, your main nighttime conductor, appears on stage as darkness sets. It has many jobs, but one of them is to signal *leptin* (the "eat less" hormone) to dial down, since you don't want to be hungry during sleep. Leptin is an important player in the symphony orchestra, yet you want it to play loudly only when the conductor tells it to, and rest when it's supposed to.

Think of deep sleep as an hour and a half during which you are "zoned out" or "in another world" because of the beautiful music occurring inside your brain. As in a symphony hall, when the music starts, the admission doors close so the guests can enjoy the peacefulness of the performance. When that performance is disrupted, your brain does not enjoy such beautiful music.

SLEEP DISRUPTORS

You've heard of "hormone disruptors"—chemicals we eat that sabotage hormone performance. There are also "chemicals" that can disrupt sleep, such as:

- Airway obstruction (see page 263)
- Caffeine (too much or too late; see page 178)
- Energy drinks
- Acidic and spicy beverages, such as decaf coffee or carbonated beverages (acidity can aggravate reflux)
- Alcohol (see page 298)
- Medications (see page 280)
- Artificial food colorings (see page 74)
- Going to bed upset or angry (see page 188)
- Bedtime TV (see page 185)
- Reading your email/social media (can trigger stress)
- Paying bills (can raise stress hormones)

HOW SLEEP CHANGES AS YOU AGE

The challenge to get a good night's sleep grows as we get older.

Lighter sleep. As we transition from adolescence to midlife, the amount of deep sleep we get lessens. At forty, age has already taken away 60–70 percent of the deep sleep you were enjoying as a teenager. By the time you reach seventy, you've lost 80–90 percent of your youthful deep sleep. In other words, the older you are, the lighter you sleep. (This could be another explanation for why "losing your memory" is a usual side effect of aging: deep sleep is your memory-bank fertilizer, so the less deep sleep you get, the worse your memory becomes.)

Earlier sleep. The next change that occurs as we age is our circadian rhythm gets pushed back. Our melatonin peaks earlier, prompting us to go to bed earlier. This explains why many seniors "hit the wall" by 9 PM, when those neurochemical nags prompt them to retire. Early to bed and early to rise is truly what fits the sleep neurochemistry of the senior brain.

Disrupted sleep. Age also brings more sleep *fragmentation*, often caused by weakening of the bladder, and waking to deal with it. Sleep fragmentation translates into decreased sleep efficiency, meaning the number of hours you are actually asleep while lying in bed lessens. Add to this the senior sleep nuisances of arthritis and other ailments of aging, and the sleep-disruption side effects of many medications to treat those ailments.

Sleep is key to a healthy brain; it's when your brain (and muscle) growth fertilizer, human growth hormone, peaks. Yet, as you can see, the older we get, the harder it is to meet the quantity and quality of sleep we need. It becomes fragmented, shallower, and less refreshing. This mismatch makes it even more crucial that older readers follow the sleep recipe we present in this chapter.

The shorter your sleep, the shorter your life.

Dr. Sleep Well

OUR SLEEPY SEVEN RULES FOR FORMULATING YOUR PERSONAL SLEEP RECIPE

While your sleep at night is as individual as your personality by day, here are the top science-based tips that help most people get a good night's sleep.

1. EAT IT

Learn when to eat to get a good night's sleep. For most smart sleepers, the earlier you eat, the better you sleep. Finish dinner at least *three hours* before bedtime to allow your stomach to empty. You may want to enjoy a light snooze snack sixty to eighty minutes before falling asleep, since it takes an hour or so for sleep-well nutrients to reach your brain. However, eating too much too close to bedtime sets up your stomach for night-waking reflux.

The earlier your meal, the better you feel.

Dr. Sleep Well

Snooze foods. What you eat before bed can make a difference in your sleep quality, too. The snooze effect of various foods can be very individual. For most people proteins perk up the brain and are better eaten for breakfast, while carbs are sedating and are preferred for dinner. Keep your own snooze-food journal to find your personal snooze foods. One of Dr. Bill's favorites is an egg.

Sleep-friendly foods—ones high in slow-release carbs and sleep-inducing nutrients, such as tryptophan, calcium, and magnesium—include:

- Dairy products: yogurt, cottage cheese, cheese, milk
- Eggs
- Soy products: soy milk, tofu, edamame
- Seafood
- Game meat: elk, venison
- Poultry
- Cherries
- Greens: spinach, kale
- Beans
- Lentils, chickpeas
- Hazelnuts, peanuts, walnuts
- Sesame seeds, sunflower seeds

Sleep-disrupting foods are those that are too spicy or too acidic and can trigger reflux and heartburn.

CLEARING CAFFEINE CONFUSION

Neurologically, caffeine is known as a "psychoactive stimulant." It works by blocking receptors in the brain for adenosine, a neurochemical sleep prompt. Adenosine starts building in the body as soon as you are awake, and when it reaches a certain high concentration, it becomes a sleep-inducing neurochemical. By blocking these receptors, caffeine prevents adenosine from saying to your brain, "Go to sleep!"

Will a cup or two of java in the morning harm your sleep at night? It depends. Some people are more *caffeine sensitive* than others, meaning enzymes in their livers are less effective at metabolizing, or "using up," the caffeine. And we become more caffeine sensitive as we age. But even for the average person, caffeine's effects last roughly five to seven hours, so drinking a couple cups of coffee, even late in the morning, confuses the brain. It reduces not only the quantity of your sleep, but also, because you are ignoring your internal sleep prompts, the quality of your deep sleep. If you don't wish to ditch your morning java, at least keep in mind that—like how the earlier your meal, the better you'll feel— the earlier you consume caffeine, the better you are likely to sleep.

2. MOVE IT

When you move your body a lot during the day, come nightfall your biochemical sleep prompts are stronger. Exercise increases the neurochemical nag adenosine, which prompts the brain to sleep. Researchers have found that both your quantity and quality of sleep are better the more active you are during the day; exercisers tend to have both an increased amount of deep sleep and an increase in total sleep time. This makes sense; the more active you are during the day, the more you need to rest and refresh at night.

Want to get even better sleep? Back to mom's advice, "Go outside and play." Sleep scientists reason that this is because the sunlight you get while exercising outdoors further revs up your wakefulness (or alertness) hormones, prompting your brain to need more rest at night.

However, it's best not to exercise vigorously within three hours prior to bedtime. The best sleep-inducing effects result when you exercise four to six hours beforehand, for example, by enjoying a pre-dinner workout around 5 PM.

BREATHE BETTER, SLEEP BETTER

Since the brain is a high user of oxygen, any habit—like exercise—that improves delivery of oxygen and blood supply to the brain is smart. In our medical experience, breathing problems while sleeping, called obstructive sleep apnea (OSA), is one of the most frequently overlooked contributors to brain-health problems, including the "D"s, such as ADHD.

When the brain doesn't get enough oxygen, it gets stressed and releases stress hormones into the bloodstream—the opposite of what is meant to happen during sleep. These excess stress hormones travel throughout your body, and instead of you healing while sleeping, you start hurting. Blood vessels narrow, your heart rate speeds up, your blood pressure rises, and instead of resting, your heart overworks. Your hypothalamus dials down thyroid-stimulating hormone, leading to hypothyroidism, a common cause of feeling "tired." High cortisol blocks the action of the natural anti-diuretic hormone, vasopressin, and you wake up to pee several times, further disrupting your sleep. Your blood sugar stays high, causing wear on your sugar-regulating system that, over time, leads to type II diabetes. Eventually, stress from the brain not getting enough oxygen at night also leads to *adrenal fatigue*, or feeling "tired and blah"—when the adrenal gland says, "I'm exhausted! I'm literally burnt out, so I can't produce enough of your energy hormones, such as cortisol and DHEA."

If you have OSA, beware of substances, such as alcohol, sedatives, and narcotics, that can make it worse. A helpful resource to explain how OSA interferes with sleep is *Sleep Interrupted*, by Steven Park. (See related topic, "Liram's story," on how OSA can be a hidden cause of ADHD, page 263.)

3. THINK IT

Happy thoughts promote happy sleep. At no other time in the twenty-four-hour day does the advice "don't worry, be happy" have the biggest effect. As you're drifting off, don't stew on the day's toxic thoughts. Remember, *if you stew, you stir.* As you shut your eyes, shut your mind to any thoughts that could keep you awake. Recite a prayer, your favorite poem, or replay a happy scene from that day or the past. Dr. Bill's nightly go-to-sleep thought is from one of his favorite movies, *White Christmas*, as Bing Crosby sings, ". . . and I fall asleep counting my blessings."

COACH HAYDEN NOTES . . .

My son often struggles with anxious thoughts before bed. As a way to soothe his mind, I prompt him to talk about what he wants to dream about. This helps him calmly fall asleep to happy thoughts.

PICTURE YOUR HAPPY PATH TO SLEEP: INSIDE DR. BILL'S SLEEP SANCTUARY

Martha and I line the trail to our bedroom with fun photos of our family adventures. A second happy dose of pictures adorns the walls of the bedroom itself. As we stroll this path, it helps put us in a happy *frame* of mind to enter our sleep sanctuary.

Because your visual center takes up a big chunk of cerebral real estate, happy scenes trigger happy thoughts that calm the brain. As you nestle down to sleep, imagine your brain prompting, "Only happy thoughts allowed in here."

MODERN SLEEP AIDS

We clever, sleepless humans have a history of coming up with techie solutions to make it easier to fall asleep and easier to stay that way. See an updated list of our favorites at AskDrSears.com/SleepSolutions.

4. SET IT

Inside your brain is an internal clock, called the suprachiasmatic nucleus (SCN). Also called your *circadian clock,* this brain clock is located near your optic nerves, which register the amount of light coming into your brain. It's set to prompt you to go to sleep as it gets dark, and wake you up when it gets light. When darkness hits, your internal clock sends a neurochemical text message to its brain buddy, the pea-sized pineal gland, to secrete melatonin, appropriately known as the *darkness hormone.* The light of sunrise then dials down these sleepy neurochemicals and dials up the wake-up-and-work prompt, cortisol. Instead of setting an "alarm," your brain sets an "alert," naturally awakening based on your hormonal "get up" prompts. This means less stress on the brain and you wake up more refreshed.

Notice your sleep/wake clock is naturally set to sleep when it's dark, between 9 and 11 PM, when the two sleep prompts, melatonin and adenosine, are the highest.

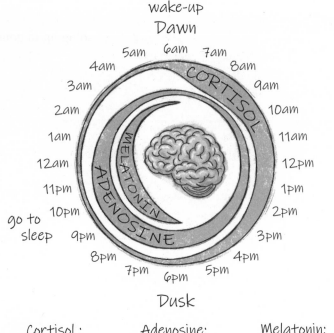

wake-up
Dawn

Dusk

Cortisol:
get up and go

Adenosine:
go to sleep...

Melatonin:
really go to sleep

MIND YOUR MELATONIN

"Sleepyhead" has a neurochemical basis. Mother Nature, our favorite sleep doctor, provides a neurohormonal prompt called melatonin. Melatonin gives us the heads-up that we soon need to put our heads down to sleep, and is one of the most healing and happiness-promoting medicines your nighttime pharmacy makes. If it's not disrupted, melatonin begins its neurochemical nagging around dusk or 6 PM, peaks between 10 PM and 4 AM (earlier in infants, toddlers, and seniors; later in teens), and reaches its low point around sunrise. (Its sleep-prompting buddy adenosine works similarly.) When you follow our sleepy seven rules for sleep, melatonin is dispensed right on time to help you get a good night's sleep.

Have a set bedtime routine. One great way to help set your brain clock is to establish a series of habits you do every night leading up to going to sleep. In one of Dr. Bill's earliest books, *The Baby Sleep Book*, he wrote about the importance of a bedtime routine that sets up a child's brain to think: "After I go through all of these steps, sleep is expected to soon follow." Our tired adult brains can also be programmed to drift off to sleep better following a *set bedtime routine*.

A bedroom routine Dr. Bill and Martha do and teach:

- Dim lights an hour before bedtime and turn computer, TV, and other screens off.
- Enjoy our peaceful picture path to the bedroom (see page 180).
- Brush teeth.
- Take a warm shower or bath.
- Offer a prayer of thanks for blessings.
- Enter our nest.

Avoid alarms. *How you awaken* is also important for setting your circadian sleep/awake clock, not to mention the difference it makes in how you start your day. Sunlight or bright lights wake you by turning on your happy hormones, serotonin and dopamine. But consider what goes on in your body when you "alarm" your sleeping brain. You're soundly asleep, probably in the REM dreaming state, when a loud sound startles your brain into releasing stress hormones, which mistakenly tell your body, "Emergency out there, run for your life!" You're already stressed and you're not even out of bed.

Avoid using old-fashioned alarm clocks to wake up in the morning. If you must awaken at a set time, try using one of the new wake-up gadgets that use your favorite gradually-increasing-in-volume sounds, music, or light to ease you into awakening. You may be surprised to discover, after you master your own sleep recipe, that most mornings you naturally awaken around the same time.

WAKE UP REFRESHED

Two more tips for starting your day the brighter way:

"Stay and stew" is bad for you. Don't stay in bed and dwell on the problems you may face during the day or lament the stresses of the previous day. Instead, start your day with good thoughts, similar to the ones you enjoyed while drifting off to sleep. Remember, *stay and stew is bad for you.* Upon awakening, quickly shift your mind into an attitude of gratitude, giving thanks for a restful night's sleep. Immediately after easing out of bed, look out the window or walk outside, lift your arms to the sky in windmill fashion, and end with your hands prayerfully together in thanksgiving. You're "preloading" your brain for a happy day, programming it with "This is how I want to think and behave the rest of the day." The brain listens and obliges.

Drink up when you wake up. Your brain doesn't like to start the day dehydrated from all the water you breathed out (and sometimes peed out) during the night. Soon after awakening, down two to three glasses of water.

5. FEEL IT/OBEY IT

The two neurochemical sleep nags, melatonin and adenosine, will tell you when to go to sleep. But you have listen to them. Pay attention to when they start making you feel sleepy. They will help you determine your *personal smartest bedtime.*

Don't fight sleep. A simple summary of why many sleep poorly is their clocks are out of sync. Your own internal clock is when your body tells you it's time to go to sleep and time to wake up. The external clock, the standard for everyone, is sunrise and sunset. Getting a good night's sleep requires these two clocks to be in sync. While modern living pressures often keep that from happening, the more you can get your internal clock and the external clock to match, the more you will enjoy the neurochemical benefits of all the stages of sleep, and the more refreshed you will feel when you wake up. Get into the habit of getting into bed just before the time that your sleep neurochemicals usually prompt you. For example, say you usually feel the strongest urge to go to bed around 10 PM. Get in the habit of starting your before-bed ritual by 9:45 so you can be ready to drift off to sleep when the sleep nags peak. Again, missing that window of sleep-quality opportunity by ignoring these nags and waiting until 11 PM is a setup for a poor night's sleep.

JOURNAL YOUR PERSONAL SLEEP RECIPE

On mornings you wake up refreshed, and on those you didn't, journal the differences:

- Did you listen to your natural sleep nag or not?
- Which snooze foods and snooze snacks worked and which didn't?
- What other sleep aids have you found to work?

---

6. DIM IT

The happy phrase "light up your life" is true—during the day. But at night our brains are genetically programmed to dial down in darkness. They thrive on the rhythms of natural light, including a gradual dimming during the evening, from daylight to moonlight and starlight. On the contrary, our brains do not sleep well when lit up with high doses of artificial light in those golden few hours before bedtime.

> ## COACH HAYDEN NOTES . . .
>
> I have noticed I get irritated at night when I'm tired and surrounded by bright lights. It's as if my brain is having to work extra hard. As soon as I dim the lights, I feel a huge sense of relief.

The reason for this is *artificial light delays the delivery of melatonin*. When the sun goes down, the brain is programmed to release melatonin to increase its appetite for sleep. When lights come on instead, the brain gets confused. Even the average lighting throughout the house and in your bedroom can delay *50 percent* of the melatonin released into your brain. Fluorescent lights are even worse. In other words, the more artificial and lighter the light, the later the melatonin release. Some of our patients have a sleep-better habit of wearing light-dimming glasses a few hours before bedtime. Think *the brighter your nightlife, the shorter your sleep life*. This melatonin delay is the main reason we can have trouble getting to sleep.

To better prepare the brain for sleep, install dimmers throughout the house. If you have automatic dimming on your smart phone, program it to turn on in the evening. *Dimming off to sleep helps you drift off to sleep.* It's best to have lights out at least an hour before bedtime.

Twinkle, twinkle, little house. Dr. Bill's longtime friend and typist Tracee shared this suggestion for dialing down at night: "Our family turns off bright lights and turns on soft 'twinkle lights' throughout the house when the sun goes down. Not only are they pretty and fun to look at, but they're a subtle reminder to the family that it's time to settle down and enjoy quiet activities before bedtime. This sets the tone for a calm and restful night's sleep."

Get the LED out. The more artificial and lighter the light, the later the melatonin release. The worst offenders are *LED lights* (light-emitting *d*iodes). LED lights have an economic and environmental advantage over older light bulbs because they last longer and use less energy. But like the modern "food processing" that enables us to eat more cheaply (but makes medical care more costly), this type of artificial lighting could be more expensive for the brain. The light wavelength of blue LEDs, as they are called, particularly dials down melatonin release.

Screen out screens. Most adults, and now teens and even younger kids, also look at LED and other brightly lit screens within an hour before bedtime—the absolute worst time of the day to be looking at artificial light. Studies have shown that using electronics with blue LED light within two hours prior to bedtime reduced melatonin by more than 20 percent.

Sleep scientists have known for more than twenty years that screen time before bedtime not only shortens sleep hours but also disrupts sleep quality. If you must indulge in screen time before bedtime:

- Watch a "feel good" movie with a happy ending, to dial down daily stress and dial up calming hormones.
- Avoid scary or upsetting stuff, which can trigger sleep-disturbing stress hormones.
- Especially if you have an LED screen, wear blue-light-blocker glasses, which can delay the melatonin suppression effect of bright lights.

BOOK IT!

To read or not to read before bedtime? The answer depends not only on what you read, but also how the words you read are lit. An interesting study found that volunteers who read a book on an electronic tablet for several hours before bed had a *50 percent delayed melatonin release* compared with those who read a print book before bed. Electronic reading *delayed melatonin release for up to three hours*, and e-book readers took longer to fall asleep compared with print book readers. Yes, e-readers can read themselves to sleep, but they are more likely to remain awake. The electronic reading gets more sleepless: electronic readers got less REM sleep and felt less rested the following day. They also suffered melatonin suppression for several days after the study.

7. COOL IT

The cooler your sleep sanctuary, the sounder you sleep. For most people, a temperature between 60°F and 68°F is best. Let's take another trip inside your brain to see why. Residing next to your brain clock is a tiny cluster of thermometer-like neurons in the hypothalamus that sense your body temperature. This smart thermometer texts your nearby brain clock (the SCN): "It's cooler in here, so the sun must be down. It must be nighttime." The pineal gland then receives this text message and responds by dispensing melatonin.

Take a hot bath. As an added sleep-inducer, take a hot shower or bath before going to bed. Contrary to what you may think, this will actually cool your body. Warm water dilates the blood vessels in your skin. Dilated blood vessels are your heat radiators—or, in this case, heat dissipaters.

As you step out of the shower, don't be too quick to dry off. Let the water evaporate, which also has a cooling effect. Then hit the "sack." By the way, the word "sack" is a great description of the optimum bedding situation. Loose-fitting sheets allow more body heat to escape, keeping you cool as you sleep.

Cool your mind, too. "Cool it" applies to your state of mind, not just your body temperature. *Never go to bed angry.* Do not ruminate. If you've had a difficult day, distract yourself first with a peaceful scene, a song you love, or a romantic moment.

Dr. Bill uses a sink-side notebook to write down thoughts or the next day's agenda before he brushes his teeth, rather than worrying about it as he's drifting off to sleep.

EASE YOURSELF TO SLEEP

Don't "try hard" to go to sleep, which often backfires by raising your anxiety level and delaying your drift-off. Instead, "try *soft*": let sleep naturally overtake you by habitually sticking to your personal sleep recipe. Most sleepers don't quickly "fall" asleep. Instead, they gradually slip into sleep over the course of ten to twenty minutes.

This *sleepy seven* recipe works for most people most of the time. Try it, refine it, personalize it, and enjoy a good night's sleep.

DESIGNING YOUR SLEEP SANCTUARY

You will spend more time in your "nighttime pharmacy" than any one single place. Make it fit your personal sleep recipe. To love your bedroom and look forward to entering your nighttime nest, try our tips:

- Don't dread your bed. Pick out the perfect bed for you.
- Position the head of your bed near an open window, so you'll get the fresh air, which your breathing passages should prefer.
- Use dimmers (automatic, if possible), candlelight, and blackout curtains to keep your room dim before bed. Remember, the older you get, the darker your bedroom needs to be.
- Install an automatic thermostat that dials down to around 66–68°F in the evenings.
- Turn off electronics, or put a bit of black tape over any monitor lights.

- Turn on "white noise," such as a fan.
- Enjoy allergy-free carpeting, blankets, and pillows.
- Enjoy a cup of chamomile tea.
- Enjoy a sniff of bedside lavender oil. Dr. Bill likes to put a drop just beneath each nostril, then take a deep breath and imagine a relaxing scene, such as lying on the beach listening to the waves roll in and out or resting in a hammock next to a babbling brook.

OUR SLEEPY SEVEN KEYS TO A NICE NAP

In addition to opening your sleep pharmacy at nighttime, you can get a quick, small dose of refreshing "medicine" during a twenty-minute nap. You may have heard this referred to as a "power nap." Personally, we prefer "a pause that refreshes." (Decades ago this was the slogan of a high-caffeine cola. Our "treat" is much better for you!) Nappers refresh their brains to think and learn better for the rest of the day.

Here are our seven keys to taking a nice nap.

1. **Plan it.** When possible, reserve a twenty-minute "timeout" from your busy schedule. Schedule this at the same time each day. Just as your brain likes a set bedtime, the brain likes a set nap time.
2. **Feel it.** Feel your natural nap prompts. You may feel a "nudge" to nap during the day that's similar to but gentler than the neurochemical prompts you feel at bedtime. Most people feel these little nudges to nap around eight hours after awakening. If you wake at 6 AM, your optimal nap time is probably around 2 PM.
3. **Time it.** A twenty-minute "catnap" is usually the right amount of time to get a bit refreshed. Longer naps (more than

thirty minutes) may cause grogginess because you drift into deeper stages of sleep, from which the brain doesn't like being awakened. However, as with nighttime, experiment with your personal nap recipe. If you naturally nap longer than thirty or even fifty minutes, and you awaken feeling refreshed, you experienced a full sleep cycle and that may be your perfect nap. The most refreshing nap length is usually somewhere between twenty and fifty minutes. Your nap length is how long you were actually asleep, not how long you laid down, because it may have taken ten to fifteen minutes to actually fall into the light sleep of a nap.

4. **Routine it.** Get in the habit of drifting off with a routine similar to your nighttime pre-sleep practice. Your brain likes routine, and will more easily drift into a nap following the thoughts or images or actions you've programmed it to follow at night.

5. **Darken and silence it.** Try your best to duplicate your bedroom setting, your sleep sanctuary. Use earplugs, mute your cell phone, close the doors, quiet the room, turn out the lights, and draw the shades. Just closing your eyes may still let too much daylight through, especially if napping outside, so try "nap shades," like sunglasses.

6. **Position it.** Napping while reclining upright at a 45- to 60-degree angle is usually better than lying flat, especially if you've had lunch within the past hour. Sleeping too long and too soon after a meal is likely to trigger gut pains such as reflux or indigestion; your gut metabolism slows during sleep, and lying flat prevents gravity from helping move the food through your digestive system. A reclining chair with a foot rest makes for a more comfortable nap.

7. **Brighten it (after).** Upon naturally awakening, get up, walk around, and look out the window. Even better, walk outside. These prompts tell the brain that naptime is over and it's time to get back to work.

ADDITIONAL SLEEP RESOURCES

As you are designing your personal sleep recipe, these trusted resources will provide helpful tips:

- soundsleepsolutions.com
- National Sleep Foundation (sleepfoundation.org)
- American Academy of Sleep Medicine (AASM.org)
- *Why We Sleep: Unlocking the Power of Sleep and Dreams*, by Matthew Walker
- *Good Night: The Sleep Doctor's 4-Week Program to Better Sleep and Better Health*, by Michael Breus, PhD

A TALE OF TWO SLEEPERS: SLEEPLESS SAM AND SLEEPY SUZY

Sleepy Suzy follows our sleepy seven recipe and wakes up every morning with brain and body refreshed. Sleepless Sam doesn't.

Sam sits too much during the day. Because his body isn't as tired at night, his brain gets less of a prompt to sleep. The neurochemical nags he does get, he often ignores. He also has erratic bedtimes and awakening times, which confuses his brain. It doesn't know when to dial down and when to dial up.

Sam lights up his life at night, just the opposite of what his brain wants. For him, nighttime is time for catching up on emails and for other screen time. He habitually checks his email "one last time" around 10 PM, the worst possible time to expose his sleepy brain to artificial light.

To add more insult to sleep injury, he overdoses on caffeine, both too much and too late in the day. He eats too much too late, and goes to bed with his stomach too full. His gut brain protests by refluxing, giving Sam heartburn, which further keeps him from a restful night's sleep.

Sam is restless in bed, tossing and turning, for at least an hour before he finally gets to sleep. He wakes up several times at night, so his sleep is fragmented. He doesn't receive many of the medicines his brain and body need to thrive. And when he finally falls into a much-needed therapeutic sleep at 6 AM, he receives a high dose of stress hormones when his alarm goes off. Sam awakens literally alarmed, unrefreshed, and unhealed.

Which sleeper do you think lives healthier, happier, and longer? Sleepy Suzy or Sleepless Sam?

PART II

HOW TO USE THE SMART TOOLS TO HEAL

In Part I, you learned the self-help skills to make brain health your hobby—the goal of this book. In Part II, you will learn how to put these skills into practice to help you heal from what you have. Beginning with anxiety and depression, the top debilitating "brain stuff" ailments at all ages, and moving on to ADHD, neurodegenerative diseases like Alzheimer's, and even alcohol disease, these chapters will help you build your *personalized brain-health recipe*—including, when necessary, how to smartly and safely partner these skills with prescription pills.

As you read through these next chapters, upgrade from a *reader* to a *doer* by referring to "Your Daily-Do List"—our checklist of the most important skills to enjoy a calmer, happier, and smarter you—in the Appendix, page 339.

CHAPTER 7

<p style="text-align:center">~~~~~</p>

HOW TO HEAL FROM ANXIETY AND DEPRESSION

"Welcome," Dr. Vince says, opening the door to his office to admit Sadie, his next patient. "I'm so glad you came in today. So many depressed people suffer in silence. Believe me, I know how devastating and debilitating depression can be. For more than fifty years I've helped many patients, and my family members, heal. I can help you. Tell me what's going on. How are the emotions you don't like feeling affecting your daily life?"

"Doctor," Sadie replies, taking a seat, "I just want to feel more joy in my life. Sometimes I go back and forth from feeling high to low. I have no interest in romance. I have trouble getting out of bed in the morning and some days I just feel flat. I'm late to meetings at work and appointments with my friends and family—sometimes I don't even show up! I just don't feel motivated. I keep putting things off, even though I've been meaning to do them, but I just can't. I don't exercise and my diet is a disaster . . ."

Dr. Vince nods. "Sadie, you're stuck! You're in brain lock. You know what skills will help you feel better, you just can't make yourself do them."

"Yes!" Sadie confirms.

"How is your sleep?"

"My sleep sucks! In fact, some days my whole life sucks. I lay in bed for hours but sleep just doesn't come. And when I finally do fall asleep, I wake up constantly throughout the night."

"What's the first thing you do when you wake up in the morning?" Dr. Vince asks, delving deeper.

"I stay in bed, sometimes just not wanting to get up and face the day. But the longer I'm in bed, the worse I feel. I know I should get out of bed, but again, some days I just can't. I feel the dread right now as I tell you this!"

"How's your eating?"

"It goes up and down with my moods. Some days I just don't feel like eating and have no appetite. On really stressful days I overeat and crave sweets."

"That's not unusual! When our brains are stressed out, we have diminished crave control," Dr. Vince explains. "Sometimes when your brain hurts, so does your body. Are there any other illnesses you want to tell me about?"

"Well, I've had a flare-up of my thyroiditis. I'm getting sick more often than I used to, and it took me twice as long as my friends to get over the flu. My gut feels queasy a lot, too. Some days I get blocked up and constipated; other days I have diarrhea."

"Do you suffer from high blood pressure or high heart rate?"

"My doctor says my blood pressure is okay, but sometimes I can feel my heart racing."

"What used to give you joy in your life, Sadie?" Dr. Vince asks.

Sadie sighs. "I used to enjoy playing the piano and tennis, but I've stopped doing either. My biggest problem really is that, some days, I just can't make myself do what I know I'm supposed to do—much less the things I used to enjoy. And some days I just dwell constantly on past mistakes, even small ones, which makes my self-worth tank even more."

"Sadie, you need a lift," Dr. Vince says.

Sadie begins to reply, but starts crying instead. "Sorry, Doctor!"

"It's okay, Sadie! Crying can be healing. You're going through a lot! If I were to take a spoonful of your tears and analyze them, they would probably have a high level of your excess stress hormones. Sadie, has anyone ever told you, 'You're so sensitive'?"

"Yes! How did you know?"

"It's common with people who experience depression, and anxiety, too. Celebrate your sensitivity. It's a gift. Sensitive people are what make the world go around. They become caring and compassionate, because they have an innate sense of justice, of right and wrong, and tend to be easily bothered by what insensitive people think of as normal. School bullies and terrorists suffer from insensitivity. Things that should bother them don't.

"Yet sensitive people need to carry around a bigger toolbox to keep their sensitivity from dialing up too high and making them overly or incessantly bothered—that's anxiety—or dialing down too low and making them depressed and flat. So we're going to help you set your sensitivity dial just right."

"I like that! But, Doctor, what's really going on inside my brain? It's not just that I'm sensitive, right? Why does my brain feel so bad sometimes?"

"Simply put, it could be an overanxious amygdala or an unhappy hippocampus that triggers your anxiety and depression. Yet it's inaccurate and unscientific to pinpoint the cause of all mood imbalances to only a few specific areas of your brain.

"Perhaps in another session I will take pictures of what's going on inside your brain, using neuroimaging. We may even be able to measure exactly which of your neurotransmitters are out of balance. These in-depth looks into your brain can help us prescribe the right pills—*if* you end up needing them. But we'll start by giving you a few self-help *skills* that we know can help all areas of your brain, and therefore address whatever and wherever your imbalance is."

"That makes sense," Sadie says, starting to smile.

Depression and its close cousin, anxiety, are two of the most debilitating diseases of the decade. They often occur together as "mood disorders" and are the top causes of mental unwellness.

What causes them is complex, so the best way to treat them is, too. Because all brains are different, all neuro-chaos is different, as individual as DNA. It would be unscientific and overly simplistic to say all depression is caused by "this" neurochemical imbalance, and therefore "this" treatment will help balance it. Still, most mood imbalances are caused by *some* type of neurochemical imbalance. And the self-help skills we introduce in this chapter—based on our more than fifty years of experience as doctors, coaches, mates, and parents—are the ones we've seen heal most people most of the time.

We've divided this chapter into three parts. First, we take you inside the brain to explain the disharmony of its inner workings when you feel anxious or depressed. Second, we help you work out your step-by-step *personal peace recipe* to bring you back into cerebral harmony. Third, we talk about the smart use of mood-mellowing medications (pills) when the self-help skills need extra help.

Get ready to feel happier!

DEPRESSION, ANXIETY, AND YOUR CEREBRAL SYMPHONY

Recall our analogy in the first section of the book: the brain is like the greatest symphony orchestra that ever played. The players are our neurohormones; when the music they play is in harmony, we have emotional wellness, and when the music is out of harmony, so are our moods.

For your cerebral symphony orchestra to play in hormonal harmony, your brain must have *biochemical balance*. Let's meet some of your players:

- *Cortisol, adrenaline, and norepinephrine*: These *play loud and fast* (think drummers and horns). Their job is to excite the brain, revving us up and making us more alert.

- *GABA, acetylcholine, and serotonin*: These *play softly* (think violins and flutes), dialing down the loudness (called "neural noise"), mellowing us out, and helping us "cool it."
- *Dopamine and oxytocin*: These *create melodies* of joyful rhythm in our cerebral symphony, invoking both calmness and pleasure.

When these cerebral symphony players are in hormonal harmony, your brain is in *biochemical balance*. When they are not, the cerebral orchestra is suffering from a chemical *im*balance.

When your "loud and fast" hormones play too loud for too long, your symphony orchestra suffers from anxiety. But when they don't play when needed, or play out of key, the music becomes flat and unmelodious, resulting in depression.

Neuroscientific discoveries have taken much of the mystery out of mental illness. Just as physical illnesses are often the result of a biochemical imbalance, so are most mental illnesses. However, unlike with many physical illnesses, we don't yet have widely available laboratory tests to measure which neurochemicals are out of balance, nor can we measure how medications change the levels of these neurochemicals. This basic challenge of treating mental illness is why both patient and doctor have to work smartly together to determine the right brain-balancing program for each individual.

Yet the good news is that, generally, both depression and anxiety are preventable and treatable when you use the self-help tools you will now learn.

ANXIETY AND DEPRESSION: WHAT'S GOING ON IN THERE?

Anxiety is a form of fear-related stress. Besides its emotional and behavioral manifestations, like feeling fearful and emotionally on edge, anxiety may cause physical changes, like muscle tightening. You may act aggressively or retreat. Emotionally, you may cry or feel overwhelmed.

Anxiety can be good or bad. When we experience anxiety at the right level and at the right time, called "adaptive anxiety," it can—like other kinds of stress—be our friend. It's when we allow it to overtake our mind, by remaining too high for too long, that it becomes hurtful, or "pathological anxiety."

Depression is a feeling of sadness, loss, and helplessness. It is the inability—no matter how hard you try—to muster up joyful and mood-elevating neurochemicals. The conductor (you) wants to play beautiful music, but the neurochemical players don't oblige.

There are two major types of depression: *exogenous* and *endogenous*. Exogenous (*from the outside*) depression comes from our experiencing a serious loss, such as of a loved one. Often such a loss strikes so unexpectedly and imbalances our neurochemicals so abruptly that we are temporarily unable to rebalance our neurotransmitters. This type of depression benefits from the short-term use of pills, which lift up our emotions long enough to help us muster the skills to rebalance our brain.

Endogenous (*from the inside*) depression usually results from an ongoing imbalance of our neurotransmitters, such as dopamine, serotonin, and adrenaline. This type of depression is where the skills in our healthy brain plan really shine and are more likely to lessen your need for pills.

Remember, the root cause of mood imbalances such as depression and anxiety is neurotransmitter imbalance. And that imbalance is the prime target of our healthy brain plan.

WHY NEUROTRANSMITTERS BECOME IMBALANCED

One cause of chemical imbalances in mood disorders is that the level of our neurotransmitters (or NTs) is too low. When our brains are in balance, the production of NTs is sufficient for our emotional and physical well-being. When excessive anxiety or other factors depress our supply of NTs, you get depressed, too.

Adrenal fatigue. We often call endogenous depression "adrenal fatigue," and it can result from being chronically overanxious.

Suppose you are trying to cope with stressful events, such as difficulties at work or home, or medical or emotional problems. When you experience stress, and think stressful thoughts, it triggers the release of stress hormones, namely cortisol (which is usually highest in the morning and helps prompt your body to get active) and adrenaline (which provides energy for you to work hard during the day).

These stress hormones are largely made in the adrenal glands (so named because they rest on top of the kidneys, or ad-renal). A constant bombardment of biochemical text messages from stressful thoughts literally wears out, or stresses out, the adrenals, causing what's known as *adrenal fatigue*. As a result, these energy-providing glands no longer produce enough cortisol, so you generally feel "down"—you lack energy, motivation, and creativity. So, yes, you have a "chemical imbalance," but it's both in the brain (stress hormones too high) and the body (energy-producing hormones too low).

The reason depressed patients often have trouble getting out of bed in the first place, and even more trouble doing what they need to do, is that cortisol, their get-up-and-go hormone, is out of balance. Thanks to their stress hormones being too high for too long, their energy-producing hormones are too low at the time of the day (daytime) when they need them the most.

A dip in dopamine. Depressed patients also suffer from something we call a *dopamine dip*, where their dopamine levels are too low for too long. Dopamine is the "zest for life" neurotransmitter, your pleasure-and-reward hormone. Without it, you don't feel pleasure. This is why depressed people often resort to drugs such as cocaine, amphetamines, and marijuana to boost their dopamine and release them from anxiety.

When your NTs are low, it's as if you've run out of gas. And the best way to refuel is to use skills to increase NT production and bring them back into balance.

For many people, their NTs are out of balance because their lives are out of balance. Our healthy brain plan skills help you raise and balance your NT levels naturally. For example, the skills you learned in part I are designed to prevent severe dopamine dips, especially going outside and playing (see "Movement in Nature: Exercise Squared," page 113), isometric exercises ("Isometrics: The Anytime, Anywhere Smart Exercise That Anyone Can Do," page 111), deep breathing ("Take a Deep Cleansing Breath," page 141), and building a grateful brain ("Have an Attitude of Gratitude," page 142, and "The Grateful Brain," page 143). In our own family lives and medical practices, we have noticed that even the *anticipation* of a joyful event that is happening soon—"Going on vacation next week!"—seems to dial up our dopamine.

What "lifts" you up and what settles you down become your *personal peace recipe*. You may notice an immediate lift just looking out the window on awakening or after a brisk walk. Journal what lifts you up and make these daily ingredients part of your personal peace recipe.

Once you understand these concepts, from here on you won't say "I'm feeling depressed" or "I'm feeling better," but "I'm NT depleted" or "I'm in NT balance."

"BIPOLAR" AND THE BALANCED BRAIN

Some of the most productive and successful people—those who are gifted with the two "Cs," creativity and compassion—are also "bipolar." We don't want you to be so out of balance that you can't use your creativity and compassion in productive ways. But we also don't want you to become less creative or less compassionate (the side effects of many pills, which work by flattening your moods). Many people learn how to ride their personal emotional rollercoaster wisely, happily, and to their advantage—dialing up or down when needed, and sometimes just stopping the ride for a while to let the engine rest—and that's what we want you to learn in this chapter.

BECOMING RESILIENT

Our healthy brain plan, especially the stressbuster tools you learned in chapter five, helps you be resilient—it gives you the ability to bounce back from life's setbacks. How resilient you are determines whether you sink into depression or are able to turn a problem into an opportunity for growth when a stressor or downer appears.

WHY YOU GET ANXIOUS

Anxiety disorders are the most common mental illnesses in the US, affecting 40 million adults, according to the National Institute of Mental Health. More than 350 million have anxiety disorders worldwide; they're much more common than depression. And while many people with anxiety disorders go undiagnosed, and still others misuse alcohol, illegal drugs, or over-the-counter medications in harmful attempts to self-medicate, prescriptions for anti-anxiety pills tripled between 1996 and 2013.

Anxiety is one of the leading causes of insomnia and other sleep disorders. And an anxious mind is an anxious heart: a person with anxiety is four times as likely to develop high blood pressure, and is twice as likely to have a heart attack.

But anxiety isn't all bad. The same neurohormones that cause anxiety are also in charge of revving us up when we need to run faster, compete better, or meet a deadline. This is why it can be a challenge for neuroscientists and healthcare providers to treat anxiety. You want to dial down a patient's excess anxiety, but not so far that they become less creative and less motivated—and what skills and pills are needed to achieve that balance can be very different for different people.

Meet your anxiety neurotransmitters. There are three main neurotransmitters involved in anxiety: norepinephrine, serotonin, and GABA. Norepinephrine triggers the anxiety response, while serotonin and GABA balance and quiet an overactive stress response. When these three are out of balance, it causes communication issues between the brain's rational center and its emotional center. When your emotional center is overactive, and the rational area of your brain is unprepared to put its normal checks and balances on the emotional brain, anxiety turns on.

Your automatic control system is on overdrive. Another way to understand anxiety is by taking a deeper look at our autonomic nervous system, the ANS. (Read more about your ANS, page 121.) The get-up-and-go or "become anxious" feeling is prompted by our SNS releasing the neurohormones epinephrine and norepinephrine, which are more easily remembered by calling them *adrenaline*. When you get anxious, these neurohormones stay on overdrive.

Your frontal lobe has lost control. The part of the brain that makes us human is our cerebral cortex—the rational thinking center in the top layer of our brain, especially the frontal lobe. This higher-up brain center can control, for better or worse, the more primitive fear centers down deep in the middle of our brains. Our healthy brain plan skills work by training these higher centers to control the overactivity of the lower ones. We "feel anxious" when these higher centers are unable to control the fear centers. Think-changing your brain helps convince the higher centers to take control by telling the lower centers, "Hey, you down there! There's nothing to be afraid of. Your fear is unrealistic. Now cool it."

THE ANXIOUS BODY

Anxiety may begin in the mind, yet because of the body–mind connection you learned about on page 130, if the mind is out of balance, so is the body. Anxiety can cause a pounding heart, sleep problems, muscle tension, and even itching! Long-term anxiety increases the risk of hypertension, cardiovascular disease, and stroke, to name only a few.

HIGH-PERFORMANCE ANXIETY

"The brighter the better" is not entirely true. The same brain biochemistry and brain centers that drive creativity in high-performance people—such as doctors, CEOs, inventors, and actors—also makes them more prone to anxiety. Their high-performance centers dial up higher, but they also sometimes have more difficulty dialing down when the need to perform at that higher level is over. A study done by one of our favorite authors, Dr. Nancy Andreasen, who wrote *The Broken Brain*, showed that people who are the brightest and most creative also tend to have a higher incidence of depression and anxiety.

This performance–stress connection also explains why it's important for high performers to be doubly careful about over-medicating, lest it lead to them becoming underproductive. Better a double dose of skills and a half dose of pills.

"OH, JUST GIVE ME A PILL!"

Sitting in Dr. Vince's office, her tears dried, Sadie confesses, "Doctor, I not only struggle with depression and anxiety, I also struggle with whether or not I should take medicines. When I first made this appointment, I just wanted a pill. I just wanted to feel better fast. Yet I've seen how pills have made some of my friends feel numb, even though they've helped others."

"*Struggle* says it all," says Dr. Vince. "You're not only struggling to balance your brain, you're also struggling to find out what medical tools—therapy and medicines—are right for you. I'm here to help you figure that out.

"Sadie, I'm a skills and pills doctor. Sometimes you need both pills and skills, sometimes only skills to start off with, but I never recommend starting with pills without skills. We are going to go step by step, stocking your toolbox with self-help skills on the way to your personal peace recipe."

YOUR STEP-BY-STEP PLAN TO ALLEVIATE ANXIETY AND DEPRESSION

Dr. Vince continues, "Remember, Sadie, that we previously discussed how the root cause of mood imbalances is neurotransmitter imbalances, which cause your cerebral symphony to play less-than-joyful music. The tools you will now learn will help you regain your NT balance so your brain can again play beautiful music."

"Wow!" Sadie says. "Let's start!"

AN IMPORTANT NOTE FROM DR. VINCE

While the following plan will help all people, those with severe depression—which prevents sleeping normally and interferes with your job and/or caring for children—should seek a physician's attention, especially as it can require possible use of medication.

STEP ONE: PUT YOUR HAPPY CENTER BACK IN CONTROL OF YOUR SAD CENTER

Reread the sections on these centers, especially how your rational and emotional centers need balance (page 205). Also, reread "How to Think-Change Your Brain from That of a Negative Person to a Positive Person" (page 132), from chapter five.

Many people who dial up too high (anxiety) are also prone to dialing down their neurohormones too low (depression). So, these tools not only keep you from remaining dialed up too high, they also help lift you up when depressed. Because your amygdala, if out of balance, becomes your anxiety center, doing some anti-anxiety exercises that calm your amygdala and perk up your happy centers should help.

Tame the trigger. As soon as you feel an anxious thought coming on, dial down your amygdala and dial up your rational "cool it" center. Here are three ways to do that:

1. **Mind-mute, trash bin, and *think this, not that***. Within seconds of a toxic thought infecting your peace of mind ("What if I lose my job . . ."), nip it quickly before it has time to "sink in." Think "mind-mute" (blank out your mind for a few seconds). Next, think "trash bin" as you dump the worry. Finally, quickly think about or visually replay a scene from your "happy file" of good memories instead. (See page 134 for more on dealing with toxic thoughts.)

2. **Deep breathing while humming**. We discussed deep breathing techniques for stress relief in chapter five, but they're also a great tool for treating anxiety and depression. Taking deep in-and-out breaths, with a prolonged hum (see technique, page 141), quickly dials down your worry centers and dials up your happy centers by raising your happy hormones. Cortisol and adrenaline fall, and dopamine rises. While deep-breathing, think *I'm dialing up my dopamine*. Within seconds of deep breathing, you can change your brain from the chemistry of anxiety to the chemistry of calm.

3. **See it, hear it**. If you're still struggling to achieve "peace of mind," add sight and hearing to further distract you from your anxious and depressing thoughts. As you look at a serene sight, such as the sky, trees, or other nature scenes, add sounds, such as a mantra, to your breathing ("feeeel gooood, loooove moooom"). The energy-hungry visual and auditory centers of the brain will suck up the mental energy you otherwise would have wasted on worrying.

Oftentimes, if my usual "don't worry, be happy" tools don't
quickly bring me peace of mind, I open my cellphone gallery
of happy scenes. I call them "my lift." They help me to reroute
my sad and pity-party thoughts into a happier path.
–New mother Erin Sears Basile, coauthor of *The Dr. Sears T5 Wellness Plan*

When a worry thought starts . . .

- Mute your mind, trash-bin the thought, and *think this, not that*.
- Breathe deeply.
- See and hear happy sights and sounds.

People with anxiety and depression often feel that their anxious thoughts control them instead of them controlling their thoughts. These personal-peace-recipe tools can put such people back in control. Remember, happy thoughts increase blood flow to your happy centers, which stimulates the brain to grow those centers. That's TCB (think-changing your brain) in a nutshell.

Practice makes perfect. Your mind loves happy habits. At first you may feel you are working hard to use all these mood-switching tools. Eventually, they will become routine. Automatic negative thoughts will quickly be replaced with automatic positive thoughts.

Sleep anxiety away. Normally, cortisol—the neurohormone that, in excess, causes anxiety—dials down to its lowest level during sleep. In anxious people, it stays high. (Read all about how to lower your nighttime stress in chapter six.)

COACH HAYDEN NOTES . . .

When I'm struggling with depression, I crave sleep. In fact, it can feel like sleep is all I'm capable of in the moment. When I'm able to give my brain the sleep "medicine" it is demanding, it thanks me by feeling better the next day.

Spend time in nature. In nature therapy studies, neuroscientists have noticed that people most prone to anxiety were also the ones that benefited most from being outdoors. That makes sense: Anxious people have their arousal centers dialed up too high, but nature can dial them back down.

LET THE SUNSHINE IN

The more darkness in your life during the day, the more darkness is likely to enter your mind. Neuroscientists believe that the reason depressed people feel so much better outdoors in sunlight is because sunlight dials down the stress-stimulating hormone (called the corticoid-releasing hormone). Imagine yourself looking out the window or going outside on even a cloudy day and picture the dials on your stress-hormone factory dialing down.

THE DEPRESSION-INFECTION CONNECTION

Ever wonder why you feel more "down" when you have an infection, even just a cold? The reason getting sick in your body can make your mind upset is related to the discovery that, as we learned on page 28, one of the root causes of depression and most other mood imbalances is *neuroinflammation.*

When you get sick, your immune system produces *inflammatory markers*: biochemical soldiers called cytokines that also act like biochemical text messages, telling your body to turn up the infection fighters. Research has shown that people suffering from anxiety and depression tend to have higher levels of inflammatory markers circulating in their bloodstream. Also, this connection is why autoimmune illnesses, such as arthritis and lupus, tend to flare up with depression. Unfortunately, these cytokines can also penetrate the otherwise protective blood–brain barrier and disrupt the firing of your happy hormones, such as serotonin.

This is another reason why reading—and doing—our healthy brain plan in part one is preventive medicine against depression and anxiety: our plan helps balance these biochemical infection fighters so they're strong enough to fight, but not so strong that they disrupt the chemical balance of your brain.

NATURE'S ANTIDEPRESSANTS

What you eat can make a big difference in how you feel, because the foods you eat have a big impact on *neuroinflammation*, the root cause of mood disorders. Therefore, when suffering from anxiety and depression, it's especially important to eat more foods that quell neuroinflammation—the foods we recommend in chapters two and three.

Science lists many nutrients that, if you don't eat enough of them, can contribute to depression, including B_{12}, selenium, zinc, folate, and many more. But the many compelling studies on one of them, omega-3s, have led the American Psychiatric Association to recommend adding *at least 1,000 milligrams* per day of omega-3 fats to the overall treatment regimen for mood disorders. Omega-3s not only boost happy hormones, they also strengthen the hippocampus—one of the areas of the brain that (as shown in PET scans) shrinks in the severely depressed.

Neuroscientists at the Department of Nutritional and Lifestyle Psychiatry at Massachusetts General Hospital showed that people who ate less of a standard American diet (SAD) full of processed foods and more of a Mediterranean-style diet rich in real foods, vegetables, and seafood (similar to what we recommend in chapter two) suffered less depression. The SMILES Study revealed similar antidepressant effects of this diet. These new discoveries in the field of nutritional neuropsychiatry upgrade the historic Hippocrates mantra of *Food is medicine* to the modern mantra of *Food is medicine for your mind*.

STEP TWO: TRY TALK THERAPY

In addition to following our healthy brain plan, you may be helped further by a few sessions of *cognitive behavioral therapy* (CBT), also discussed on page 130, which we prefer to call *how to think-change your brain*. CBT helps you learn how to *reject and replace*: reject the negative, for-no-reason worry, and replace it with "that's the real me" thoughts. We doctors like to call this "reframing." (See "Framing Difference Positively," page 259, for more about how it helps anxiety and depression.)

HOW TO GET THE BEST FROM A THERAPIST

To get the best out of your time in therapy, ponder these pointers:

- *Go forward.* While sometimes it is necessary, and healing, to dig up past issues that influence present problems, be smart in not dwelling on the past, for these reasons:
    - If you can't change it, don't dwell on it—that's Mental Healing 101.
    - The further back and the deeper you dig, the less accurate the memories. Some people, especially if *prompted* by "Do you remember any awful, scary scenes?" remember events in ways that are exaggerated, unreliable, and even fabricated. "Tell me about your mother" may not be fair to your mother or helpful to you.

    In other words, *it's okay to dig, but not to dwell.*

- *Go prepared.* Try the advice in chapter five on stress management. Then you can tell your therapist what works for you and what doesn't, and they can help fill in the gaps.

- *Be honest.* As you are telling your story, your therapist is trying to gauge just how much your "stuff" affects parts of your life, such as your marriage, parenting, work, and so on. If some days you feel life is not worth living, or you are having similar thoughts, let your therapist know so they can better judge the depth of therapy you need.

GET A SKILLS COACH

Self-help skills can be easier talked about than done when you are depressed and feeling stuck—you know what skills to do and why you should do them, but you just can't. Here is where a so-called skills coach is helpful. They literally come and usher you out of bed, take you for a walk or a swim, and watch funny movies with you, even when you make excuses that you "don't want to" or "can't." Your skill coach could be a caring friend, a spouse, or anyone who holds you accountable to use the skills you learned in part I.

STEP THREE: GET A PERSONALIZED MEDICAL CHECKUP

Balancing your body to balance your brain is Mental Wellness 101. That's why treating depression and anxiety requires thinking outside the box—or, in this case, outside the brain.

The most common mistake most brain-health doctors make is *focusing only on the brain.* Brain quirks, like most illnesses, result from a chemical imbalance, and the greatest challenge of brain medicine is to smartly investigate just what chemicals are imbalanced and in which organ systems the imbalance is occurring. Remember, your brain is biochemically connected to all other systems—and especially the cardiovascular, gastrointestinal, immune, and hormonal systems. So before we can fix your brain imbalance, we must be sure these four systems are in balance.

MORE MONEY FOR THE MIND

While mental illness is now the number-one medical problem in the country, sadly it is the one that insurance companies reimburse least often. Translation: Despite being the top medical illness, it often gets the least investment when it comes to care.

At a recent brain conference Dr. Bill attended, a prominent psychiatrist shocked the audience by revealing what really goes on behind the scenes at medical practices owned and operated by management companies, whose directors know little about medicine but a lot about money. She was working as a psychiatrist for a large medical corporation and trying to give her mind-needy patients her best. But she was reprimanded by the CEO for taking too much time with her patients. Her orders: "You're allowed twenty minutes per patient. Just write a prescription for Prozac and move on." Fortunately, she did move on—to a personalized medical practice that enables her to care for patients in the ways they need.

Get a personalized medicine consultation. "Personalized medicine" is the name of the five-star medical approach that delves more deeply into your personal habits and needs in order to see what changes you need to make for optimal health. Doctors who use this approach may also advertise as specialists in "integrative medicine" or "functional medicine." Because you are unlikely to find one doctor who is a specialist in all four systems, start with one generalist and get a complete medical checkup and blood analysis for:

- *Cardiovascular health markers.* Remember, the better your blood flow, the better your brain.

- *Inflammatory markers.* Most brain chemistry imbalance has its root cause in neuroinflammation.
- *Hormonal balance.* This should include a gender-specific hormone panel and a thyroid panel (thyroid imbalances are a big cause and contributor to brain imbalances), as well as tests for your levels of stress hormones, blood sugar, blood insulin, and other hormones that affect your brain health.
- *Gut balance.* Fix your gut to fix your brain. Read how to do this in chapter three. For updates, see AskDrSears.com/microbiome-testing.
- *Toxic metal levels.* Testing your urine may detect these neurotoxins, which can cause severe long-term problems such as dementia. (See related story, page 318.)

Often your personalized medicine specialist will refer you to, or consult with, a specialist in one or more of these five areas.

THE BRAIN AND THYROID—PARTNERS IN HEALTH

Many brain-health problems, such as depression and anxiety, have roots in thyroid health problems. If thyroid hormones get dialed down too low, you can become depressed, obese, and mentally slow. If thyroid hormones become too high for too long, you grow anxious, thin, and agitated. You want the right balance of hormones in your thyroid just as you do your brain. Remember, while your brain is your body's largest endocrine organ, it also communicates with its buddies, like the thyroid gland.

Fill in your gaps. The first step toward better mental health—before we think about brain pills—is to correct any imbalances that your medical checkup finds. Suppose your personalized medicine specialist finds your thyroid hormone levels are low, and after you start taking the appropriate hormone, you feel better. You then continue taking the pills and doing the skills to help your ills—and hold off on other mind-altering pills for the time being.

If your body biochemistry is all in balance, great! Then you can start directly addressing your brain biochemical imbalance with the confidence that there isn't another source of imbalance going on in your body.

BEHAVIORAL RISK FACTORS FOR DEPRESSION AND ANXIETY

The genes you inherit play a minor role in depression and/or anxiety, but the growing science of epigenetics teaches us that your genes are not your destiny. While you may inherit a "tendency" toward depression and/or anxiety from your parents or grandparents, whether or not this tendency "switches on" and you actually suffer from depression and/or anxiety is under your control. While you can't change your genes, you can change how they behave by the way you behave—by changing the following risk factors, which scientific research has shown to be present in people with the highest incidence of mood disorders:

- More time spent indoors
- More time spent sitting
- Eating a high junk-carb diet
- Eating a low smart-fat diet
- Eating more animal-based and fewer plant-based foods
- Eating more artificial food additives
- Excess time spent on social media
- Excess time spent in social isolation

All of these are behaviors that are within our control. Change your behavior, and you can change your brain. That's what our healthy brain plan is all about. Review the illustration on page 22 showing how your brain and major organs are *interconnected*.

ADDING PILLS TO SKILLS

To treat mood disorders, we strongly recommend a skills-before-pills model. This doesn't mean never using pills, but it does mean holding off on pills until we see whether a skills-only approach will work.

Pills work by balancing your neurotransmitters—which can arguably be considered the biochemical basis of "you." So when you mess with "you" by using medications, you do change the person—sometimes for the better, sometimes for the worse. Psychiatrists notice this in patients on medication. While some become less depressed and less anxious, many also become less creative, less interesting, or "flat." As one of Dr. Bill's patients said to his wife, "You are no longer you."

SKILLS AND PILLS—SOMETIMES PERFECT PARTNERS FOR BRAIN HEALTH

Doctors are always searching for the reason or the "why" behind the effects we observe when a patient responds to a combination of pills and skills. Our common sense tells us that it's for several reasons. Because exercise increases blood flow to the brain, perhaps this helps the pills go deeper to affect more areas of the brain. Also, the most common side effects of pills are weight gain, tiredness, and loss of sexual interest. Exercise prevents the first and perks up the other two. Finally, "prescribing" skills along with pills and holding patients accountable for not only "Did you take your medicine?" but also "Did you *do* your medicine?" helps patients feel that they are in control, not the pills—that they are still the chief conductor of their cerebral symphony orchestra, even when an outside conductor (the pills) is brought in to help.

An additional reason we prefer to start with skills is that there are three significant challenges in prescribing pills to treat chemical imbalances in the brain:

1. Recent studies have shown that certain chemicals—including medications such as SSRIs (see "How Pills Work," page 221)—are depleted from the body through liver metabolism at a rate that varies from person to person. (Your physician may be able to test how enzymes affect these medications in your system.)

2. We also cannot measure the level of neurochemicals in your nerve gaps. (Yes, there is a "gap" in this reasoning behind most psychopharmacology use; see "Nourishing Your Neurotransmitters," page 9.) So we have to guess the "right dose for you" based on the "average dose" in studies. Your brain is not "average."

3. Suppose you get lucky and feel better within a few weeks of taking the drug. You and your doctor agree: "It's working, keep taking it." But eventually the brain may outsmart the drug, in a process called downregulation, and you are left in what we call "the hole": having to take higher doses with more unpleasant effects, or to add more drugs to treat the side effects.

We want to help you avoid both of these common, depressing scenarios: increasing drug doses that lead to unpleasant side effects, and you becoming less "you."

PILL POWER AND WILLPOWER

The overfocus on pharmacology to heal your mind often takes the responsibility away from the individual and their self-help skills. The right dose in the right proportion of pills and skills is the key to lasting healing.

Medication may help you regain chemical balance, but here's a very important point we want you to understand: the medication should never ever replace "you." Brain medications are designed to be part of your toolbox, which you may carry around for a short time or a long time, but they must always be a smaller part of your toolbox than the "real you," meaning your self-help skills. What we don't want is for the "new you" to be mainly made by the medicines you take instead of the medicines you make.

Using a sailing analogy, you must always be at the helm. People around you, and even medications, may help you trim the sails better, get out of overwhelming winds, and so on, but you must always be the one at the helm. When the self-help tools that you learn finally prevail and balance the pills, then you're more likely to thrive.

SMART AND SAFE USE OF PRESCRIPTION PILLS

Many people with depression in particular get "stuck." They know all the skills they should be doing, but they just can't get going. Often a short course of the right medication may be just what they need to jumpstart their skill-doing. Then, once the effects of the skills click in, they can usually, but *gradually*, wean off the pills.

In the meantime, matching the biochemical imbalance drug with the missing biochemical, and finding the right dose, is as much an art as it is a science.

When you have an infection, most doctors try to do what's called a drug–bug match. First, they culture a body fluid (such as urine, blood, or sputum) and grow the "bug" in a little dish. Then, they add various drugs to see which one kills the bug. When they find something that kills the bug, then they have a reasonable drug–bug match, and prescribe you the medication. That was simple!

Yet, for the brain, what's "bugging" you, and how to "stop" it, is not that simple. Your mood disorder does not come from outside of you the way an infection does. Your emotions are affected by many factors: how you live, eat, move, and think. So treating (a.k.a. drugging) the chemical imbalance without treating the underlying "imbalances" in lifestyle, nutrition, exercise, and attitude that contribute to it misses most of the point.

Medication can, however, be helpful—especially at first. Sometimes pills are necessary to help balance your brain so that you can practice the self-help skills that eventually reduce or eliminate the need for pills. Let's take a deeper look at why and how mood medications work, and at the best way to integrate them into your healthy brain plan, if necessary. (Also, the better you understand how mood-altering medications work, the more likely you are to take your pills as prescribed.)

How pills work. The main hypothesis attempting to explain why mood-mellowing medications work is called the *reuptake hypothesis*. This hypothesis has been popular since its Nobel Prize–winning discovery in 1970, and its mechanism has opened more doors to more drugs for more disorders.

Recall that your neurotransmitters are biochemical text messages that move from one nerve to another (see page 10). At the end of each of your nerves, there is a gap called a synapse. Picture ferryboats carrying these neurotransmitter messages from one nerve to another. When they reach another nerve, they unload the neurotransmitter. The other nerve uses it, and then the ferryboat brings the biochemical back to the original nerve's dock, storing it until it's needed again. That way, biochemicals don't have to constantly be reproduced by the nerve cell. It's one of the most energy-efficient mechanisms in the whole body. If you don't produce enough neurotransmitters, or happy hormones, such as serotonin—if your neurotransmitter levels are "depressed"—you feel depressed, because the areas of the brain that promote good feelings aren't being properly stimulated.

The reuptake hypothesis suggests that if you could increase the level of these biochemicals in the gaps between neurons by causing them to remain in the gap longer and therefore allowing nerve cells to fire longer—in other words, if you could temporarily inhibit their reuptake by the original nerve—then the level of these happy hormones and neurotransmitters would stay higher and the person would feel better. This is what most mood medications do. They work their magic in the gaps between neurons, the synapses.

Understanding this mechanism gave rise to a whole class of mood-mellowing medications called SSRIs, selective serotonin reuptake inhibitors, which theoretically increase the amount of serotonin in the gaps between neurons. The longer your natural happy hormones "fill in your gaps" at higher levels, the happier your mood. Remember, your neurotransmitters in these gaps are the language of the brain. The happier the language, the happier the brain.

Start low and go slow. Since your doctor cannot know the details of your personal "chemical imbalance," and can only make a best guess based on how you think and act, treatment should start with a low dose. If necessary, this can be increased—but only gradually, since with nearly all brain medicines the hurtful effects go up as the dose goes up.

Instead of just writing a prescription for pills, your doctor should set a goal for eventually weaning you off the pills, or at least lowering the dosage. There should be a timetable and an action plan that includes skills. This plan may take three months, or it may take years. Unlike other bodily imbalances, especially those of the heart, the brain changes slowly but smartly (see neuroplasticity and pruning, page 14). But it doesn't matter whether your plan is a fast track or a slow track. The important thing is that it is *your track*.

Remember, your doctor can't measure what's going on in your "gap." They can draw conclusions and make smart changes based only upon what you tell them. So be honest! Mental wellness providers are trained not to pass judgment on individual patients' problems. Notice your most bothersome concerns and needs, how deep they are, and most important, how they are affecting your daily living. The smartest way to help your doctor prescribe the right pill and the right dose for you is to be honest and accurate in reporting your brain changes.

Doctors are under time constraints due to the demands of HMO medicine. So in preparation for your visit, don't be afraid to write down your concerns, and then record their responses. If you don't fully understand these, ask them to clarify their answer or to recommend an article or book to help you understand your problem.

Some good news: In our medical experience, the people who healed best from depression and anxiety did so through learning the healing "medicine" of *resilience*—the ability not only to bounce back from life's setbacks, but to "bounce up" and become a stronger person through the bounce-back tools they used. They reframed their setbacks into opportunities to grow into happier and healthier people.

KRISTEN'S STORY OF HEALING FROM DEPRESSION AND ANXIETY

We want to share one of the most heart-wrenching and brain-healing recent stories from our medical practices. Read Kristen's story at AskDrSears.com/brain-health-testimonies. Two major goals of this chapter are to help you heal from anxiety and depression while giving you the self-empowerment tools to transform, as Kristen did, into a healthier and happier you.

CHAPTER 8

~~~

PREVENTING AND REVERSING ALZHEIMER'S DISEASE AND OTHER NEURODEGENERATIVE DISEASES

People who are polled about their top health worry repeat, "I'm afraid of losing my mind." Sadly, that top fear is becoming the top reality through the epidemic of Alzheimer's disease (AD). Each year nearly 10 million new patients the world over are diagnosed with dementia. In the United States, healthcare costs associated with AD approached $300 billion in 2018 alone. Eighty percent of people have a relative or family member who has AD. There are now nearly 50 million people with AD worldwide, and that number is likely to quadruple by the year 2035.

Alzheimer's is a *progressive* brain disorder in which you lose the abilities to remember memories, solve problems, and react appropriately and lovingly—basically, the brain wears out. Recently the American Academy of Neurology has recognized a new and very important precursor to AD called mild cognitive impairment (MCI), which describes the loss of function in the brain's structures that involve emotions, short-term memories, and movement.

Because all your brain centers are interconnected by biochemical text messages (your neurotransmitters), the brain tissue decay in AD that affects memory and movement gradually causes loss of brain tissue in other centers, too, such as those for smell, taste, balance, vision, and emotional control. In advanced stages, AD patients may require wheelchairs or become unable to leave their beds for the last decade of their lives.

While AD is now the number-one worry in our healthcare system, it is also one of the *most preventable* problems. New treatment plans (like the tools in chapters one through six) show that Alzheimer's disease can, to varying degrees, be stopped or reversed. Today, not only is there more *hope* for healing from AD, there are also more *tools* for doing so. There is now light at the end of the dark AD tunnel.

CLARIFYING AD TERMS

Alzheimer's is a spectrum of severity and stages. Like most diseases, doctors try to stage AD according to how it affects daily living. Terms you may hear include:

- *Mild cognitive impairment*: Early memory and emotional impairment without loss of reality. While MCI symptoms are noticeable, they have only a mild effect on daily living.
- *Dementia* (from the Latin phrase *de mentis*, meaning "out of one's mind"): An umbrella term referring to memory and emotional impairment that causes a loss of reality.

ALZHEIMER'S DISEASE: NO LONGER A MEDICAL MYSTERY

Modern neuroscientists conclude that the root cause of AD can be summed up in one word: *mismatch*.

Over the last fifty years, during which time the incidence of AD sky-rocketed, we have been forced to grow and live in a toxic environment for which we were not designed. No organ, especially one as complicated and needy as our brain, can survive a mismatch with its environment unscathed. Expose your brain to a lifetime of neurotoxic food, neurotoxic air, and neurotoxic thoughts, and it's no mystery why our brains wither with age.

Yes, "wither" is what actually happens inside the AD brain. Back to our garden analogy from chapter one. Your brain is the greatest garden ever grown, boasting trillions of "plants," or brain cells. If you don't feed, fertilize, and irrigate a garden, and keep out pests, the garden withers. Plants shrivel up. The same is true for the brain.

Inside the brain of a person with AD. AD is named after Alois Alzheimer, the insightful German psychiatrist and neuroscientist who in 1901 delved deeply into a fifty-one-year-old patient's brain at autopsy when she died following years of progressively losing her mind and memory. He found that not only was her brain shriveled up, but it was also full of sticky stuff, what he called senile plaques, and numerous tangled nerve fibers. This *cerebral atrophy* is what causes AD.

New technology has also revealed a lot about what's going on inside the brains of those living with this disability. PET scans, which measure blood flow throughout various areas of the brain, reveal two changes: AD brains don't "light up" as much as younger brains (there is less blood flow in the brain), and they have shriveled-up, or atrophied, areas.

While what happens to the brains of people with Alzheimer's is evident in autopsies and scans, even top neuroscientists remain uncertain of AD's root cause. Yes, sticky stuff, such as beta amyloid and tau-protein tangles (also known as neurofibrillary tangles), builds up in most (though not all) brains with AD. But whether this "metabolic garbage" is the cause of the disease or just a symptom (the result of a faulty garbage-disposal system in the brain) is still unclear.

SEEING IS BELIEVING

Want to view some "wow!" photographs of brain scans of people with AD? Search for "Alzheimer's" in Google's image search engine and you will see why we describe AD as a withering, shrinking brain.

Since neurologists don't presently agree on exactly what stuff inside the brain actually causes AD, here is our most scientifically accurate and reader-friendly explanation:

> **AD is a progressive buildup of sticky stuff in the blood vessels and tissue of the brain, which eventually wears the brain out.**

Our healthy brain plan helps keep sticky stuff out of brain tissue and blood vessels—exactly what we need to help prevent AD!

WHY SENIORS SUFFER FROM AD

Major contributors to Alzheimer's disease are three brain changes that usually happen as we age:

- *Blood flow lessens* (see page 18). A new term, *vascular dementia*, has recently crept into the neurologist's bag of brain illnesses. Translation: Slow and sticky blood flow.
- *The blood–brain barrier leaks* (see page 14). This allows more environmental neurotoxins to seep into the brain, where they interfere with neurochemical balance and can lead to early Alzheimer's.

- *The garbage-disposal system weakens*. This allows more "garbage," or neurofibrillary tangles, to build up in the brain. (See explanation of the glymphatic system, page 166.)

Preventing AD therefore means preventing, or delaying, these three faults.

OUR HEALTHY BRAIN AND AD PREVENTION

As mentioned in "Brain-Health Problems—Now the Top Medical Concern," despite pharmaceutical companies spending *$84 billion* developing and testing anti-Alzheimer's drugs, all have failed. Yet, as you will soon learn, while pills have failed, self-help skills have succeeded. The take-home message: When it comes to preventing and reversing AD, *skills help, pills don't*—at least for now.

This bad news is overpowered by good news: because we now know most of the contributing factors, or "triggers" (neuroscientists are not yet ready to call them "causes"), of AD, if you can tame the triggers, you can turn down your risk and severity of AD. You want that! And the best way to get it is to do—and stick to—our healthy brain program:

Brain Problems in AD	How Our Healthy Brain Plan Helps
Brain tissue shrinks	Grows brain tissue
Blood flow slows	Increases blood flow
Memory center shrinks	Grows the memory center
BGF (fertilizer for your brain garden) lessens	Increases BGF
Blood–brain barrier (BBB) leaks (see page 14)	Makes the BBB more selective and protective
Metabolic "garbage" builds up	Improves the brain's garbage disposal system
More sticky stuff is deposited	Decreases sticky stuff

Our take-home message: *Prevention* is the best anti-Alzheimer's medicine. After examining a patient with AD, Dr. Vince often thinks, *I wish I would have seen this person twenty years ago.*

> AD is a preventable disease, for which there
> is no pharmaceutical treatment.
> –Neurologist Dr. David Perlmutter

THE ALZHEIMER'S GENE

Neuroscientists have identified genes that increase a person's risk of developing Alzheimer's. Should you get genetic testing to see if you carry the Alzheimer's gene? To help make a wise decision, consider the following:

- Having a gene only tells you that you have a slightly increased risk of developing AD—not whether you actually will develop it. New insights into *epigenetics* (how your behavior and environment affect the way your genes are expressed) and *nutrigenetics* (how you can help blunt genetic tendencies by smart nutrition) are gradually giving us the reassuring message that we can take charge of our quirky genes instead of letting them take control of us. Our healthy brain plan dials down your genetic risk through clean and smart living. Our favorite analogy to understand epigenetics, and its cousin nutrigenetics, is to think of your genes as your personalized piano keyboard. You can determine the music your genetic keyboard plays by how you think, eat, and live.
- Do you really want to know? Suppose the results come back positive. Would that just increase your stress, and therefore also your risk of AD? Or:

- Would knowing change your behavior? If your results came back "negative," would you get lazy and not keep your brain as healthy? Would a "positive" result jumpstart and motivate you to get with the brain-health program?

In Dr. Vince's anti-Alzheimer's counseling, if a patient has a family history of AD, he does recommend genetic testing as an extra motivator for following our healthy brain plan.

THE EARLIER YOU START PREVENTION, THE SLOWER YOUR PROGRESSION

While you can slow the progression of Alzheimer's after you start seeing the signs (see "Ten Tools to Prevent and Reverse AD," page 235), the most effective treatment is prevention. As Dr. Vince likes to say, "Alzheimer's begins in childhood."

Prevention of AD is best begun early, in the stages when the brain is naturally changing the fastest, growing the smartest, and most open to changing. The top two stages are:

- The third trimester of pregnancy through five years of age
- Adolescence: eleven to twenty-four years of age

Following our healthy brain plan in the first part of this book during these two periods can both delay the onset and lessen the severity of AD. The younger you are when you start taking care of your brain, the better off you are. Because Alzheimer's is a lifelong, progressive disease, the earlier you start our healthy brain plan as preventive medicine, the easier AD is to prevent. To get young and unhealthy adults to get on our healthy brain plan earlier, Dr. Bill often gets their attention by telling them: "You're pre-Alzheimer's."

Alzheimer's disease progression is occurring at younger and younger ages.

ARE WOMEN MORE LIKELY TO GET ALZHEIMER'S?

Once upon a time, we thought the reason we saw more Alzheimer's in women was because women live longer. Yet more recent research shows this doesn't seem to be a major factor. While neuroscientists really don't have a definite answer, the slightly increased risk of AD in women is most likely a combination of factors, both biological and environmental.

Estrogen, a neuroprotectant, decreases in women after menopause. Also, while this is not always necessarily so, women tend to be more sensitive or easily bothered, and get more easily stressed, than do men. The sensitivity that makes a person beautifully compassionate can also make them more prone to depression and anxiety, both of which are risk factors for Alzheimer's. Add to this the fact that women are more likely to become family caregivers "sandwiched" between caring for children with "young brain stuff" and parents with "old brain stuff," and their risk of having a burned-out brain rises. While there may be some genetic factors involved, it's most likely these are female *epi*genetic factors that increase the incidence of women's slightly greater risk for Alzheimer's. (See chapter seven for how unmanaged depression and anxiety can trigger AD.)

AD CAN BE A SNEAKY, CREEPY DISEASE

Like a stealth fighter, Alzheimer's disease sneaks up on your mind, and your caring observers, slowly. Worried you might be developing AD? Its very subtle signs include:

- Mini-forgets that happen more frequently, such as "I forgot where I put my car keys" or "I forgot where I put my cell phone."
- Spending more time indoors sitting and watching TV.
- Facial expressions that become flat and disinterested.
- Not following through on plans, or just not planning anything at all.

DIAGNOSIS DENIAL

If you're worried a loved one may be in the early stages of Alzheimer's disease, be prepared to notice early red flags before the person with MCI does. Be prepared, too, for the usual denials: "I feel fine" and "I'm just getting older." People with MCI are often unaware of the early problems occurring in memory, motor, and emotional behaviors. Expect this disconnect between what you see and what the person with AD believes.

The reason that subtle memory loss (which we all get from time to time) is usually the earliest sign is that AD generally first strikes the *hippocampus,* your memory-file storage. This is because the hippocampus is the most metabolically active area of your brain. As you learned on page 19 ("Age Well: Have a Happy Hippo"), the hippocampus's high activity is a mixed blessing. On the one hand, it is what makes the hippocampus better able to self-repair and grow more new cells. On the other hand, all this hippo-activity produces more metabolic waste products (oxidants or sticky stuff), which, if not "antioxidized" by the eating and stressbuster skills you learned in part I, can quickly lead to wear and tear.

We all are going to "forget" more as we get older. What counts is how much this "normal brain aging" affects our daily living.

DR. BILL AND DR. VINCE'S "SELECTIVE FORGETFULNESS"

We both used to shine at remembering names and event dates. However, the older we get, the more those faculties are dimming. We refuse to give a scary name to it, like "memory dysfunction." Instead, we simply laugh and celebrate the memories we have and the new events we enjoy. We are also getting more easily bothered by scary scenes in movies, negative stories, and irrelevant informational overload. We prefer to celebrate our selectivity by allowing into our minds mostly scenes and stories that "feel good," quickly trashing and "forgetting" those that bother us.

A NOTE FOR CAREGIVERS

Dr. Vince's mother, Rose, began showing scary signs of AD in her nineties. One day while visiting Dr. Vince, she imagined something was after her, so she ran outside with only her nightgown on and locked the door behind her on a freezing winter day. Luckily, Dr. Vince heard her pounding on his door. He hugged her as she had hugged him as a child. When his mother became a danger to herself when living alone, such as by leaving the stove on, it was time to escort her into an extended-care facility, where she lived happily and safely until she peacefully passed at age 105.

It is heart wrenching when your mother, father, or mate begins to lose themselves and no longer recognizes you. The end-of-life journey of a person with AD can be the most stressful of any illness. Make the best memories you can with the resources you have, and say goodbye with tears and smiles.

TEN TOOLS TO PREVENT AND REVERSE ALZHEIMER'S

Remember, while aging is inevitable, AD is not. Like you learned in the section on the "belief effect" (page 156), the more hope you have (and the more you use these tools), the more likely you are to heal, at least somewhat.

Our best advice for preventing or reversing Alzheimer's is the healthy brain plan in the first part of this book.

> **Part I of this book is your anti-Alzheimer's prescription.**

However, this section includes a second helping of that advice, mainly based on Dr. Vince's experiences treating patients with his anti-Alzheimer's program, especially over the last three decades.

Besides being based on science, our ten-step plan makes sense: AD and other neurodegenerative diseases have their root cause in *neuroinflammation* and diminished blood flow to the brain. These steps improve both.

1. EAT MORE SMART FOODS, FEWER DUMB (AD-PRODUCING) FOODS

While there are many more fad "diets" than there are scientific studies to support them, nearly all neuroscientists agree that as we age we need to:

- decrease *carbs* a lot,
- increase *proteins* a little, and
- increase *smart fats* a lot.

We'd add "just eat real foods": mostly plants, and in the above proportions.

Review and follow the instructions in chapters two and three. What's good for the brain at all ages is especially good for the aging brain at risk for AD. Following this advice can help you spend your "golden years"—perhaps the last twenty years of your life—enjoying smart peace of mind.

Be especially vigilant to *avoid sugar spikes*, as people with high and spiking blood sugar levels are at much higher risk for developing Alzheimer's. Remember that spikes are a main cause of sticky stuff in the brain (see page 29). Many AD patients are put on a very low-carb ketogenic diet, and one of the reasons this diet works well for them over the long term is because it blunts sugar spikes. (See more about the ketogenic diet, page 320.)

> Do not spike your insulin!
> —Alzheimer's-prevention specialist Dale Bredenson, MD

Drink smart, too. Read chapter ten, and you will be highly motivated to follow our advice there. The less alcohol you drink, the smaller your chances of getting AD. Excess alcohol is a top trigger of AD, second only to loneliness.

SMART EATING ADVICE TO KEEP FROM "LOSING YOUR MIND" TO ALZHEIMER'S

- Balance your blood sugar (page 67).
- "Avoid spikes!" (page 21).
- Balance your omega-3s and omega-6s (page 76).
- Graze to the rule of twos (page 90).
- Eat clean and green (page 54).
- Eat smarter foods (page 30).
- Eat more smart fats, more smart protein, and fewer processed carbs (page 66).
- Eat less, chew longer (page 90).
- Eat earlier (page 176).

DR. BILL'S BRAIN-SAVING DAILY DIET

Following my monthly brain-health talks, I'm frequently asked, "Dr. Bill, what diet are you on?" My simple response is, "The real food diet," but then I see unsatisfied looks on many faces. So, I add: "While I don't count daily calories, when I eat a smart food plan"—the one you learned in chapters two and three—"my nutrient proportions come out to 40 to 50 percent smart fats, 20 to 25 percent proteins, and 20 to 30 percent healthy carbs." Also, on most days, I eat 15 to 20 percent fewer calories than I burn, to take advantage of caloric restriction's longevity-promoting perks (see explanation of caloric restriction, page 71). In other words, I eat consistent with the best advice on eating for the aging brain. (See an example of my daily eating, page 79.)

SCIENCE SAYS THESE FOODS HAVE BEEN SHOWN TO REDUCE AD

- Oily fish
- Olive oil
- Nuts
- Oranges
- Blueberries
- Cocoa
- Dark chocolate (try 80 percent cacao)
- Apple with peel (organic)*
- Beets/beetroots
- Coconut oil
- Turmeric

*To view which foods must be organic and for which foods eating organic is less necessary, see EWG.org's "Clean Fifteen" and "Dirty Dozen" lists.

SAY ALUMIN-NO TO AD

In the past decade, the "heavy metal toxin" that has received the most study as a potential contributor to AD is *aluminum*. Due to Big Chem lobbying and label loopholes, aluminum is everywhere, especially in our food, our drinking water, and many of the most common medications. Neuroscientist Dennis Crouse, PhD, in his scientifically referenced book *Prevent Alzheimer's, Autism, and Stroke*, makes a convincing case for the cause-and-effect relationship between our societal increase in added aluminum and the increase in AD.

To lower your aluminum exposure:

- Drink beverages in glass bottles rather than in aluminum cans.
- Have your drinking water tested for aluminum content and use a water filter, such as a Brita.
- Beware of certain medications that contain aluminum, such as antacids.
- Limit use of aluminum foil in food storage and cooking.
- Don't cook with aluminum cookware.

Groundbreaking science news alert! The best nutrients to prevent AD are . . . the ones we recommend in chapter two, according to a set of recent scientific findings.

Researchers from the Decision Neuroscience Laboratory at the University of Illinois took blood samples from 116 healthy people between the ages of sixty-five and seventy-five to measure their levels of the smartest nutrients, called their nutrient biomarker pattern. They also did neuroimaging studies to view how the subjects' brains looked and how much aging their brains showed. Their discoveries: The people with the healthiest brain tissue on imaging studies had the highest levels of omega-3 fats, carotenoids (fruit and vegetable nutrients), B vitamins, riboflavin, B_{12} and folate, and vitamin D, as well as the best omega-3/omega-6 balance.

The reason we were wowed by this study is because unlike most nutritional research, which relies on "questionnaire" reporting (usually a red flag for unreliability), the researchers measured blood levels to find out not only what their subjects ate, but how much of each nutrient they *absorbed*. Smart!

Again, modern neuroscience proves Dr. Mom's sage advice: eat more fruits, vegetables, and seafood.

2. GO OUTSIDE AND PLAY

Move more, sit less, and enjoy the great outdoors. Let the sunshine in. Review and do the movements in chapter four.

Movement, especially outdoors, is AD-preventive medicine that makes sense for at least three reasons:

- Reduced blood flow to the brain is a top trigger of AD. Movement increases cerebral blood flow.
- The hippocampus (memory center) shrinks in AD. Exercise grows your hippocampus.
- Exercise lowers stress hormones. Remember: Stress less, shrink your brain less. Neuroscientists recently discovered that people who regularly take their "nature pill" (a.k.a. spend more time outdoors) tend to have lower levels of the stress hormone cortisol throughout the day.

AD researchers have found that people over age sixty who exercise regularly enjoy brains that are 20 percent larger than those who do not exercise.

MORE MOTIVATION, LESS PROCRASTINATION

People with MCI and early AD often confess, "I know I need to get my butt out of bed and get moving, but I just can't." Because depression and anxiety, two unfortunate partners of AD, exhaust your energy and dampen your motivation to get going, our advice is that the sooner you feel an effect, the more motivated you are to upgrade your doing. Often, all you need to get started is one simple change, or one simple promise, such as, "I'm going to vigorously exercise ten minutes a day." When you feel the pleasant, cerebral effect of a good workout (preferably outside), some of the trash stuff is worked out of your brain and more motivation is worked into it.

FINE MOTOR MOVEMENTS ARE FINE FOR YOUR BRAIN

In some neurodegenerative diseases, the brain center that controls fine motor movements, such as playing the piano or picking up tiny objects with thumb and forefinger, gets compromised, especially by tremors. Our prescription: Use *chopsticks* instead of a fork. It's a healthy way to strengthen the fine motor movements of your fingers. If you want an extra dose, use the chopsticks in your nondominant hand.

3. GET AND STAY LEAN

Following the first two steps of this plan often lead automatically to fat loss, and the more excess belly fat you shed, the more brain tissue you keep.

Recent findings by the American Neurological Society reveal that obese people are *seven times* more likely to suffer from AD. If you are overweight (or "over*waist*"), regardless of your age, getting and staying lean is the best preventive medicine for protecting your health. The bigger your belly, the greater your risk of Alzheimer's—mainly because excess belly fat triggers neuroinflammation. (See "Bigger Belly, Smaller Brain," page 102.)

4. MEDITATE MORE, AGITATE LESS

Meditation is tailor-made for people who want to prevent or reverse AD, especially movement meditation (see how, page 153). And the good news is, seniors usually have the luxury of more time to meditate! "Stress less to keep your brain smarter" is the message you learned in chapter five. Reread it, do it, and feel it.

Also, for senior readers, read and use the anger-management tools we list in chapter five. What do you think is the age when you're most in need of "cooling it"? No, it's not teenhood, it's *seniorhood*. Researchers recently revealed that the older a person gets, especially over eighty, the more unresolved anger triggers neuroinflammation. The take-home tip: The older we get, the more anger-management tools we need.

5. SLEEP SMART

AD has increased as the number of nightly hours we sleep has decreased. Neuroscientists believe there is a correlation. People in their fifties and sixties who suffer the most disrupted sleep tend to accumulate more sticky tau-protein tangles in their brains.

As we age, our amount of deep sleep lessens. As you learned in chapter six, this is the stage of sleep that fertilizes your memory bank. This is probably why sleep disturbances are a big red flag for memory loss.

But lack of quality sleep may contribute to Alzheimer's in an even bigger way. As we said in chapter six, one of the newest insights into the cause of AD is that it may be a garbage-disposal problem rather than, or in addition to, garbage buildup. And when we miss out on the detox effect of deep sleep (see page 171), it leads to neurotoxin buildup in the brain.

Follow our logic:

- AD is probably caused by accumulation of too much "garbage," or neurotoxins, in the brain.
- During quality sleep, your brain's garbage-disposal system works more efficiently.
- Therefore, a good night's sleep could help prevent and delay the progression of AD.

Less deep sleep equals less removal of waste products; more retention of waste products equals Alzheimer's.

As you learned in chapter six, sleep is known as a nighttime cleanse, or detox, and this cleansing mechanism is highest during this stage of deep, non-REM sleep. Unfortunately, this is the stage of sleep that lessens as we age. During quality sleep, healthy *brainwashing* occurs. Buildup of neurotoxins, such as aluminum and amyloid (sticky stuff), is literally washed away somewhat during sleep. (See page 167 on how the nighttime cleansing system of the brain, the glymphatic system, increases during sleep.)

The glymphatic system is the sewage system of the brain. It works like the lymphatic system in the rest of your body, but it's composed of glia cells. During sleep is when your brain's sanitation system revs up, also known as a *nighttime cleanse*. And the cleanse is highest during deep sleep. What is even more fascinating is that during deep sleep the glymphatic system widens. Picture a street getting wider at night to allow more garbage trucks to clean up the daily mess.

Sleep better, leak less. No, it's not the bladder, it's the blood–brain barrier (BBB). As you learned on page 14, your BBB protects your vulnerable brain tissue from toxins. The weaker your BBB, eventually the weaker your brain. An interesting article in the October 2014 *Journal of Neuroscience* reported that chronic sleep deprivation (CSD), at least in mice, can cause the BBB to leak. Wow! Just another smart brain-healing effect that occurs during a good night's sleep.

BALANCE YOUR BLOOD PRESSURE

Our healthy brain plan advice—*the better your blood flow, the better your brain*—is especially important in preventing AD. A contributor to AD is *hypoperfusion* (less blood flow to the brain), and both high blood pressure and blood pressure that is too low for too long can cause brain tissue shrinkage. A consultation with a cardiologist and the tools in chapter four will help you balance your blood pressure.

6. GROW YOUR SOCIAL NETWORK

Yes, you can do that on your smartphone, but, more important, keep in real live touch with as many people as you can. In the largest study of risk factors that lead to AD, loneliness was found to be the top trigger. (See "Endnotes," "The Grant Study.")

Socializing, playing games, and group volunteering all ease you out of your worried self and into the brain-calming and mind-building effects of your social network. Physical contact is important, too. One of the questions we often ask a person suffering from early AD is "How many friends have you hugged today?" (See "The Healing Power of Touch and Hugs," page 334.)

7. BE PASSIONATE ABOUT A PROJECT

Retirement without a passion and project is a setup for AD. We need purpose to live. Purpose gives life value. And the more we value living, the more we are likely to use the brain-health tools to live smarter. Create a plan. Volunteer and serve, and your brain will get a high dose of the *helper's high*, feeling good by doing good.

8. GROW YOUR SPIRITUAL CENTER

In Dr. Vince's decades of working with patients who have varying degrees of AD, he's noticed that those with the deepest spiritual roots are not only less likely to get Alzheimer's, but also more able to both slow the progress of the disease and reverse it if they do get it. Why? Perhaps because they're more likely to be part of prayer groups, which are also involved in serving others in the community. (See "Grow Your God Center," page 158.)

9. ENJOY MUSIC–MAGICAL MEDICINE FOR THE AGING MIND

Establishing and replaying a music-memory file is a smart early investment in your brain-health retirement plan. Researchers at Massachusetts General Hospital revealed the reason why music is particularly peaceful for people with Alzheimer's: the music-memory area in the brain isn't affected by memory loss as much as other areas. This area is also linked to the amygdala and hippocampus, the emotional areas of the brain. When someone with AD listens to a song that triggers fond memories, such as a wedding dance tune, it sparks these emotional areas to release feel-good neurohormones, such as dopamine.

Make your own anti-Alzheimer's medley. Play familiar songs that trigger beautiful memories, such as a date, a wedding dance, or a fun vacation, or newer music that just elicits an "Ah, that was wonderful . . ." feeling. Dr. Bill frequently "daydreams" by diving into his bank of joyful musical and dancing memories. He plays them frequently, especially when a stressor or toxic thought infects his brain. (See related sections, "Repel 'ANTs'," page 139, and "Your Brain on Music," page 152.)

Dance Alzheimer's away. The older we get, the more we need to dance. Dancing is like taking a dose of many medicines for the mind all at once. Besides the healing touch of a good hug, you enjoy the eye-to-eye contact of a beautiful smile, the rhythm of the music, and, of course, the physical exercise. Musical movement is truly medicine for the mind.

BRUSH ALZHEIMER'S AWAY

Add good dental hygiene to your AD prevention list. Researchers who study factors that contribute to AD noticed that people who suffer the most gingivitis (inflammatory gum disease) are also more likely to get AD.

Known as the *oral-systemic connection*, bad bacteria seep into gum tissue, usually in the crevices between tooth and gum, and infect the bloodstream, from there traveling throughout the body. This sets up inflammation, especially in the heart and brain.

To keep bacteria such as *P. gingivalis* from causing gingivitis and reaching your brain:

- Daily before bedtime, swish with warm water, brush, clean the crevices between teeth and gums with a *proxy brush* (an interdental brush), scrape your tongue, and finish off with a thorough water-flossing.
- *Brush* at least twice daily.
- *Swish* with warm water after every snack or meal.

10. CONSULT A SPECIALIST IN ALZHEIMER'S DISEASE

Think of consulting a specialist for advice as getting a preventive-medicine checkup for your mind, the way you get an annual medical checkup for your body. If you're already experiencing signs of AD, you may want to see a neurologist, specifically one who specializes in the diagnosis and reversal of Alzheimer's (and take along a spouse, friend, or caregiver!). Not only do these specialists have experience in treating this complicated illness, they also can guide you through trying modern treatments, safely and under medical supervision, such as smart supplements that are right for your brain, the ketogenic diet, and so on. Neurologists also focus not only on *what* you have, but also on *why* you have it.

An AD specialist also thinks outside the box, or in this case the brain. They often start with:

- *Cognitive testing* to see where in the AD spectrum you fit. Do you have early MCI, later MCI, or Alzheimer's disease?
- *A comprehensive medical evaluation.* In addition to a physical and neurological examination, they also do a complete blood panel especially focused on inflammatory markers and a thyroid panel.
- *An MRI, CT scan, or PET scan,* plus an *EEG,* if your exam or tests suggest it is needed.

The Alzheimer's disease specialist will also ask you, "What's in your medicine cabinet?" Many drugs seniors commonly take are the very ones that can increase the risk and severity of AD, especially the *antis*—*anti*depressants, *ant*acids, *anti*histamines, and *anti*cholinergics—and certain psychotropic medications, especially benzodiazepines ("benzos").

To get the most out of your specialist, you—if you still have both the wisdom and the medical faculties to codirect your plan—or a caregiver should bring in a bulleted list or journal listing the brain fog problems you are noticing, whether they are increasing in severity, what you have tried to reverse them, and especially how much they affect your daily life.

Centers for AD are found at most universities. To find an AD center near you, see charliefoundation.org.

PROVEN PROGRAMS TO PREVENT AND REVERSE AD

To help slow the rising tide of AD, many universities and specialists offer valuable programs. Here are a few:

DR. VINCE'S DEAR PROGRAM

Dr. Vince named the prescription for memory loss he uses with his patients the DEAR Program, and he shares it in detail in his book *The Anti-Alzheimer's Prescription*. He's happy to report that the statistics and outcomes of those who have gone through his program have been outstanding, especially since many of his patients' families had previously been told, "Go home, there is nothing to be done for your loved one." Of the first hundred patients he has seen:

- 41 percent improved
- 32 percent stabilized
- Only 27 percent continued to deteriorate

A critical factor was the time of intervention. Those with early signs of dementia (scoring 5–9 out of 10 on their mental status exam, where

10 is the highest) did the best. Those who scored 0 to 2 on the test had less improvement, yet many still benefited.

DEAR stands for **D**iet, **E**xercise, **A**ccentuate the Brain's Reserve, and **R**est and **R**ecovery.

Diet

The DEAR diet—basically the Mediterranean diet, and similar to the one we outlined in chapter two—is not a cure for AD, but rather a way to prevent it from happening or to delay its progression. We know the Mediterranean diet helps both the heart and the brain, mainly because it's a diet high in seafood, such as salmon and anchovies. Balancing your blood sugar in particular is one of our top "prevent AD" goals.

In the DEAR diet, those who followed the smart eating advice you learned in chapters two and three, and the smart drinking advice you'll learn in chapter ten, are the ones who showed the least progression and the most reversal. The earlier a person starts DEAR eating, the more likely they are to not get Alzheimer's in the first place.

Exercise

The movements we advise in chapter four, especially isometrics (page 111), are some of the best anti-Alzheimer's preventive medicines. To watch Dr. Vince doing his exercise program, see AskDrSears.com/DrVince.

Accentuate the Brain's Reserve

We neurologists teach the concept of "cognitive reserve," which refers to your brain's resistance to damage and its huge capacity for growth, maintenance, and repair.

"Protect your cognitive reserve" is one of the top preventive medicines I prescribe, especially to young adults. The more you grow and nurture your cognitive reserve in the early decades of life using the brain-growth tools you learned in part I, the less your chances of getting AD later in life.

But even as a senior, engaging in challenging games, new experiences, and other "mental exercise" is a great way to increase your cognitive reserve. It's not only physical exercisers who have larger and brighter brains; *mental* exercisers do, too. As we saw in chapter one, smartness means making the right connections in the brain. Mental exercises help your nerve fibers make smarter and more lasting connections—giving you deep, strong roots that will help prevent AD.

Rest and Recovery
Use the sleep-well strategies you learned in chapter six, and the stress-management tools you learned in chapter five to rest and recover effectively.

DR. BREDENSON'S ALZHEIMER'S PREVENTION AND REVERSAL PROGRAM
If you Google this program from Dr. Dale Bredenson, you can read about the hundred people who went through his program, many of whom showed dramatic improvements, including growth of the hippocampus (the brain's memory center) and improvement in cognitive tests.

Dr. Bredenson's program, like many other successful anti-Alzheimer's programs, includes an assessment of the health of every organ, blood tests for a personalized chemistry profile, and other outside-the-brain tests.

THE MIND PROTOCOL FROM DR. TIETELBAUM
In Dr. Jacob Tietelbaum's program, MIND stands for optimizing **M**etabolism, such as hormonal balance and insulin sensitivity; clearing up **I**nfections; practicing smart **N**utrition; and not doing **D**rugs that can be harmful to the brain. Dr. Tietelbaum's program also looks for and treats underlying metabolic problems, such as thyroid imbalance, that aren't really "Alzheimer's" related.

DON'T BE A SCARY STATISTIC

If the current trend continues, by age eighty-five, *50 percent* of seniors will have AD. OMG! That's the most startling statistic we've ever heard.

More hopefully, however, 50 percent may *not* get AD. Which half do you want to join?

People who practice the tools in our healthy brain plan are less likely to suffer AD. Those who don't are at high risk of losing some or much of their cognitive abilities. It's that simple! And you're never too young to start protecting your brain health.

VIEW IT. DO IT!

See a video of Dr. Vince explaining how to prevent Alzheimer's disease at AskDrSears.com/Alzheimers.

CHAPTER 9

~~~~~~

ADHD: CELEBRATING THE "DIFFERENCE," HEALING THE "DISORDER"

I n 1999, Dr. Bill and Martha had the honor of being on *Good Morning America* to discuss their newest book, *The ADD Book*. Our host, Joan Lunden, asked Martha, "Where did you get the material for your book?" Smiling, Martha responded, on national television, "It's my husband's autobiography!"

ADHD* (attention deficit hyperactivity disorder) has become an increasingly common diagnosis for people of all ages. What you're about to read comes from our nearly fifty years of experience in counseling kids and adults with this label—frequently given incorrectly—and from Dr. Bill's growing up with ADHD before this "D" label was invented. Also, we

---

* In this chapter we use the term *ADHD* to also include *ADD*, but some people (more commonly, girls) have what we call "silent ADD," where their problem is more one of focus than hyperactivity or impulsive behavior.

owe much of what we have learned and write about to our smart mentors, many of whom may merit some brain-difference label. You will meet a few of our "influencers" in this chapter to see how they used their special gift to their advantage. And, thanks to new technology and a deeper understanding of the many overlooked triggers of ADHD, you will learn about effective non-drug approaches to managing ADHD for both children and adults.

While some of our discussion focuses on ADHD in children, the brain-wiring differences also apply to adults with ADHD. And most of the self-help skills, and smart use of pills, apply to all ages.

SCARY STATISTICS

- Between 2009 and 2019, the number of children who were labeled as having ADD or ADHD more than doubled and the use of prescription medicines to treat ADHD tripled. Approximately 10 percent of school-age children are now taking prescription medicines for their ADHD.
- The diagnosis, or label, of "ADHD" is being given at earlier ages.
- Eighty percent of all ADHD prescriptions worldwide are issued in the United States.
- Schools are more concerned—and confused—than ever before on how to handle kids with ADHD and, quite frankly, don't have the time, energy, or resources to do so.
- Teachers are often quick to recommend children be medicated, which is why at least once a week we see a family in consultation for "the teacher thinks Johnny has ADHD."
- For children who *do* have ADHD, one reason why early diagnosis, guidance, and treatment is so important is

that adolescents with ADHD, especially inadequately treated ones, are at increased risk for poor educational outcomes, teen pregnancy, substance abuse, and motor vehicle accidents.

- Experts who deal with ADHD estimate that, of children who are diagnosed as having ADHD, around 50 percent will have varying degrees of symptoms as adults.

Your family does not have to be part of these scary statistics.

ADHD IS A "DIFFERENCE," NOT A "DISORDER"

Imagine you are coming into Dr. Bill's office for consultation about your child because you or their school think they may have ADHD, or you think you may have ADHD yourself. Here are a few items to keep in mind during the consultation.

ADHD is a *difference*, not a *disorder*. This is the first and most important point to understand. It is often a *description* of the person's learning, behavior, and focus challenges, rather than a diagnosis of something that is wrong with them. Yet, depending on one's living, schooling, and work environments and on their LEAN factors (*l*ifestyle, *e*xercise, *a*ttitude, and *n*utrition; see "Our Healthy Brain Plan at a Glance," page 3), it can become debilitating.

> **ADHD is a *description*, not a *diagnosis*.**

There have been many famous ADHD-ers. Thomas Edison, light bulb inventor and lifelong innovator, had the typical profile of someone with ADHD. One of his biographers wrote of Edison's brief school experience, "He alternated between letting his mind travel to distant places and putting his body in perpetual motion in his seat."

Winston Churchill, who was prone to hyperactivity and impulsivity, channeled his high energy into hyperfocusing on creative problem-solving that led England to survive World War II.

Finally, consider our favorite figure with ADHD, Wolfgang Amadeus Mozart. His gift of hyperfocus led him to compose operas in only a few weeks. Yet, had Mozart lived today, he probably would have been drugged for his differences, and the world would have been deprived of some of the most beautiful music ever created.

Early diagnosis and treatment are vital. If Mozart had had the right blend of childhood ADHD therapy (skills, with careful use of pills, as we'll discuss), he might have suffered less anxiety, lived longer, and, instead of dying a pauper, ended life happier and healthier, in a better social and economic position—while still sharing his music with the world. Early intervention focuses on channeling a child's uniqueness so that it can work to the child's advantage.

INSIDE THE "ADHD" BRAIN

Functional MRI scans show the brains of people with ADHD differ from those of typical people when it comes to the hypersensitivity of their amygdala. This center that governs emotions, thinking before acting, and dialing down impulsivity gets revved up more easily in some people than in others.

While brain imaging in ADHD patients does show variations in areas such as the amygdala and prefrontal cortex compared to typical people, these variations vary between individuals. This is why, for some people with ADHD, it is neuroscientifically incorrect to say, "This area in your brain is different." Yet, for others, it is correct (and reassuring!) to know that these centers function in a different way.

Many neuroscientists conclude that most of the "chemical imbalance" in people with ADHD occurs in the prefrontal cortex, where the neurochemicals for learning and attention, such as dopamine and nor-epinephrine, may be out of balance. The brain in balance behaves, learns, and focuses. A brain out of balance misbehaves, mis-learns, and can be quirky when it comes to impulse control, decision making, and general attention and behavior. (Many treatments for ADHD focus on four "core symptoms": inattention, distractibility, hyperactivity, and impulsivity.)

If ADHD is a chemical imbalance, especially of the pleasure-and-reward neurochemical dopamine, that "brain difference" would explain why people with ADHD constantly seek novel, exciting, and some-times risky experiences—reward-seeking behaviors that dial up their dopamine. It would also explain why they tune out boring subjects that don't perk up their dopamine. And it could explain why some people with ADHD just give up and withdraw from social contact (due to low dopamine), while others seek exciting and creative adventures (which increases their dopamine) and become successful entrepreneurs.

It's estimated that at least 5 percent of school-age children have vari-ous symptoms of ADHD. How many adults experience ADHD symptoms is more difficult to evaluate, but neuroscientists who study ADHD esti-mate that the number is around 3 to 5 percent. Many ADHD symptoms lessen for some teens and adults as they grow older—and wiser. And Dr. Vince and many of his neurologist colleagues note that diagnosing ADHD in older adults is more challenging because they often have other "Ds," such as depression, mild dementia, or mild cognitive impairment (MCI; see chapter eight), which can overlap and make labeling an older person with ADHD complicated and often inaccurate.

Let's dissect the label A-D-H-D to show you what it really means and why it's often misunderstood.

Attention deficit–not! Rather than having an "attention deficit," people with ADHD more accurately have "selective attention." The minds of bright and quirky kids and adults tune out facts, lessons, assignments, and jobs that seem boring or irrelevant, instead wandering to more exciting things that seem to have more relevance. For example, they may be thinking about designing a better computer, rather than a bor-ing boardroom discussion with no immediate relevance to their daily living. Yet people with ADHD can exploit a "hyper" that can work to their advantage—the ability to *hyperfocus*. CPA Joe may drift off during that board meeting, but give him a challenging tax return and he'll beat the deadline. Inattentive Alice may struggle to pay attention to a classroom history lesson, but give her the lead in a play about Joan of Arc and she shines.

Many boys with ADHD suffer from what we call *bright bored boy syndrome*. Put him in a boring classroom with teaching that he finds irrelevant and he's going to get hyper and misbehave, because the information being taught simply doesn't fit how his brain was wired. (Another confession from Dr. Bill's ADHD childhood: he remembers "playing hooky" in the second grade, faking an illness in order to stay home and finish building a bridge from his Erector Set—because he found it more relevant and interesting than what he was learning at school!) By contrast, girls with ADHD are less likely to be impulsive, disruptive, or hyperactive. Instead, their focus problems tend to manifest as "daydreaming" or seeming "spacey."

We have seen many students labeled ADHD who, once they were put into a more challenging class with a teacher that piqued their interest (usually due to smart, never-give-up parental pressure), their "ADHD" disappeared, or at least improved. The problem wasn't them. It was a mismatch between how that child's brain was wired to learn and how it was being forced to learn. While some children can survive that mismatch and still succeed, others can't, and they shouldn't be forced to.

CHANNELING THE ABILITY TO HYPERFOCUS

People with ADHD can tap into the ability to *hyperfocus* on tasks and events that they value as relevant or exciting. Parents often report how great their child's concentration is when they're playing video games. Give an adult with ADHD who loves working with numbers a page of them to organize and they're riveted to that page until it's done.

As a Little League and Special Olympics coach for more than twenty-five years, Dr. Bill has tried to place the players with ADHD as pitchers or catchers. The pitcher has to pay attention to strike out the batter. The catcher has to pay attention or he gets beaned with the ball. Yet, when you put the same child out in left field, they are truly "in left field," walking around in circles and not paying attention to the game. When Dr. Bill practiced pediatrics in Canada, he noticed that kids with ADHD also tended to be good hockey goalies, hyperfocusing to keep excellent track of the puck.

Meet Bob. Dr. Bill's friend, Bob, has been labeled "severely ADHD" and has gone through, often inappropriately, many drug trials for his *difference*. Dr. Bill had worked with him on health projects and on nutrition boards, and numerous times was asked by his loving (yet concerned and exhausted) wife to help him with his ADHD. Despite Bob's challenges, Dr. Bill noticed that he met deadlines and excelled at giving health lectures. At meetings Bob came up with creative ideas that stunned his admiring audience.

The treatment plan Dr. Bill worked out with him was to celebrate Bob's ability to hyperfocus and create an environment designed to bring out his abilities while avoiding triggering his "disorder." The changes he made, including using a standing desk to avoid the "sitting disease" (see sitting disease, page 106), getting more fresh air and natural sunlight, and eating only smart foods, made a big difference. Dr. Bill's advice to Bob that worked the best? "Walk while you talk." Bob started scheduling lengthy phone calls while he could go outside and walk in a quiet tree-lined area, which helped him focus and be even more creative.

By *channeling* instead of changing and *celebrating his uniqueness* instead of drugging the disorder, Bob has opened health and nutrition counseling services in more than fifty countries worldwide. Like many other creative and "hyperactive" people, he has made this world a healthier and more interesting place in which to live.

What "hyperactivity" really means. Now that we've explained the three terms *attention*, *deficit*, and *disorder*, let's tackle the label *hyperactivity*.

People with ADHD can struggle with being "hyperactive," or "restless" and "fidgety," especially in inappropriate situations, such as during class, at the dinner table, at parties, and at work. New insights show that when people with ADHD exhibit these behaviors, they are *self-medicating*. As you learned in chapter four, movement mellows the mind.

Movement also quiets two other features of ADHD: distractibility and impulsivity. The minds of people with hyperactivity are prone to being scattered, with many ideas popping into them at once, thanks to that hypersensitive amygdala. They often act before they think, leading to impulsive behavior.

As with selective attention and hyperfocus, these quirks can be channeled, too, through our step-by-step plan for alleviating ADHD on page 261. (See "Isometrics," page 111, for anytime, anywhere movements that can quell the need to fidget.)

WHY SOME TEENS "GROW OUT" OF THEIR ADHD

Some children with ADHD seem to "get better" when they reach their teens or young adulthood, and one of the reasons is that they have naturally pruned the plants in their brain garden.

As described in chapters one and six, your brain regularly does away with nerves that are not being used, especially during sleep, through an automatic pruning process, just as you would prune plants in a garden that were withering or not necessary. This pruning lets the brain circuits work more efficiently. It's like using your favorites list of most commonly dialed numbers versus having to sift through your contacts every time you make a call.

But while this pruning is an ongoing process, there are certain ages when our pruning becomes more efficient. During each of two "pruning ages," the first from about 7 to 8 years of age and the other from about 18 to 24 years of age, we prune *50 percent* of our neurons. Children and teens tend to be more "hyperactive," with endless electrical energy, until this pruning occurs. Afterward, these behaviors tend to settle down.

FRAMING DIFFERENCE POSITIVELY

"Reframing" is the big buzzword in ADHD counseling for patients of all ages. Early on in studying and counseling families with ADHD, Dr. Bill realized how important it was to change their mind-set from one full of negativity to one full of positivity. We all are sensitive to labels, so the language we use to describe people with ADHD matters.

Just as many parents expect to leave a doctor's office with a prescription, they also anticipate leaving with a diagnosis. So, Dr. Bill had to come up with a label for them, and decided on *quirky*. Many celebrities, executives, leaders, inventors, and comedians are quirky, and certainly, *quirky* is a nicer term than *disorder*. He still has to use the term *disorder* on insurance forms, since doctors don't get reimbursed for the diagnosis of "quirky," but while labels like that need to be in the doctor's chart, they don't need to be inside you. You have ADHD, but you are not ADHD.

This is especially important to remember when it comes to children, because if you pin a negative label on children, they are likely to act accordingly. Instead of over-focusing on the "disorder behaviors," pick out your child's strong points and beautiful behaviors, especially in the school setting. Suppose you're at a parent–teacher meeting and your child's teacher opens with, "She's so stubborn . . ." You quickly interject, "Yes, she has a persistent personality." If the teacher comments, "He blurts out in class instead of waiting his turn," quickly interject, "Yes, he is very spirited." If the teacher says, "He seems so hyper," you respond, "Yes, he sure is curious." If the teacher says, "She sure cries easily," your reframing response is, "Yes, she's very sensitive." If the teacher continues, "She gets bothered so easily," you say, "Yes, she's very compassionate." The teacher persists, "Some days he's so draining." You say, "I agree, he's very interesting and challenging."

The message you are giving to the teacher is: "Frame my child positively, celebrate his uniqueness, and channel it to work to his advantage." Framing your child positively in front of others, in front of your child, and in your own mind helps you, your child, and the people around you to celebrate your child's specialness rather than constantly dwelling on their "disorder."

Let all the derogatory terms, diagnoses, and disorders remain on the therapist's chart, the doctor's records, and the insurance billing codes—not in your mind, or your child's.

Reframe, don't blame.

TIPS FOR TEACHER TALK

A main theme of this chapter is the benefits of reframing the label "ADHD." Even calling it a "difference" may lead to people with ADHD being perceived as "less," which is why we prefer referring to it as "unique wiring." If your child's teacher has suggested they might have ADHD, your response may go something like this:

"Thank you, teacher, for picking up that my child is having some behavior and learning challenges in your classroom. We took your advice and had him evaluated by his doctor." (Opening your dialogue on a positive and respectful note encourages the teacher to listen to—and do—what you next advise.) "Our doctor strongly advised us to always *frame* our child positively, so that he doesn't feel that he's not as smart as others in the class. She also told us about exciting new studies that reveal how some growing brains, like our child's, are genetically programmed to learn differently. She suggested giving him more schoolyard exercise breaks and assigning him jobs in the classroom to help keep him focused, such as handing out papers, or giving him a part in a play."

A STEP-BY-STEP PLAN TO ALLEVIATE ADHD—AT ALL AGES

With this understanding of ADHD as a difference rather than a disorder in mind, you don't necessarily want to *change* the ADHD brain, but rather to celebrate its uniqueness while using nutrition, counseling, therapy, school matching, and so on to *channel* that uniqueness in a way that achieves these two goals:

- Channel the quirky brain onto the path for which it is wired.
- Help the person use their differences to their advantage instead of having those differences become a "disorder."

Additionally, for children:

- Instill a *love of learning*, the root of academic success.

Sometimes, when parents come to Dr. Bill's office for a consultation on their child's ADHD and ask, "Will medicine help?" he surprises them by saying, "Yes!"—then scribbling on his prescription pad (see illustration at right).

Then he goes on to describe why there is more science behind these home "medicines" helping with the "Ds"—ADHD included—than there is to support any prescription drug, as you learned in chapters two, four, and six.

Sears Family Pediatrics
26933 Estrella
Dana Point, CA 92624

Rx
- Eat smart foods ("focus foods")
- Sleep well
- Go outside and play

Refill *daily*   *William Sears*
M.D.

When Dr. Bill counsels patients with ADHD concerns, he uses the following action plan before starting—or in addition to—prescription pills.

1. JOURNAL WHAT TRIGGERS ADHD BEHAVIOR

Journal your (or your child's) behavior and learning history as far back as you can remember, and ideally from as soon as when you started having concerns. Make what you could call your "naughty" and "nice" lists: what behaviors have been detrimental and what positive personality qualities have proved beneficial. Then bullet-point:

- what *triggers* the behaviors on the naughty and nice lists.
- any changes you've made in your daily LEAN (lifestyle, exercise, attitude, and nutrition), and what results you've noticed; mainly what changes improved moods and behavior.

Once you've written down everything you can remember, continue the journal in real time, recording "good days" and "bad days," especially as you start making changes. Label the *triggers* that happened on bad days. Those triggers become the targets to start addressing right away. Bring a two-page summary of your journal to your therapist or consultant's office on your first visit, and continue to update your journal between subsequent visits.

COACH HAYDEN NOTES . . .

If they're old enough, ask your child also to journal about the changes being made. They can get creative and use pictures, drawings, or even a voice recorder. Not only can this help them process their own experience, as some of the changes can be difficult, but they might bring up something you missed. A mom once showed me a picture her six-year-old son drew when they stopped getting Happy Meals for dinner. The picture showed him going "number two." He was mad about the change at first, but he told his mom he goes every morning now, and now he runs faster. This gave her incredible insight and the opportunity to help her son see the connection between what he ate and how his body worked.

2. THINK OUTSIDE THE BRAIN: HIDDEN MEDICAL CAUSES OF ADHD

Often, when your doctor delves deeply into the health and biochemistry of someone with ADHD, they discover ailments that are the root cause of the quirky learning and behavior. *Fix the root cause* of the biochemical imbalance in the body and then see the biochemical balance in the brain.

Liram's story. When eight-year-old Liram's mother brought him to Dr. Bill for consultation about his label of ADHD, she was not only in distress about her son, but also overwhelmed with confusion from having received several different diagnoses from many doctors. Within minutes with Liram, Dr. Bill made a new diagnosis, announcing to Liram's surprised, puzzled mom: "Your son's problem is in his *mouth*, not in his brain!"

There were several clues Dr. Bill observed, all of which suggested Liram might have obstructive sleep apnea (airway obstruction that causes stop-breathing episodes during sleep). Liram was a mouth breather. During the consultation, his mouth sagged continually open. When Dr. Bill examined his mouth, he saw why: his jaw was small and his tongue was too far back. He had an anatomically narrow airway. Another clue was his "allergic salute," a white crease across the bridge of his nose caused by nasal stuffiness from allergies. When Dr. Bill looked into his nose, his nasal passages were swollen, compromising his nasal breathing. If his brain was being deprived of sufficient oxygen while sleeping, it's no wonder he couldn't pay attention the next day!

Three months after getting Liram fitted for a dental appliance to widen his jaw, decongesting his nasal passages, and teaching him sleeping positions to widen his airway (sleeping on his side rather than his back), as well as Liram practicing keeping his mouth closed while breathing, his "Ds" disappeared. (See more detail on Liram's story, in his own words, on our website: AskDrSears.com/OSA.)

Are your or your child's focusing or inattention problems due to unique brain wiring that you or they were born with? If so, focus on helping that natural brain wiring work better. But also make sure the cause of the ADHD isn't completely outside the brain—which is often the case.

The most common "causes," or triggers, of ADHD are:

- lack of smart *nutrition*.
- lack of smart *movement*.
- lack of smart *sleep*.

Following the healthy brain plan in part I to see how much difference it makes is a great place to begin.

OTHER HIDDEN CAUSES OF "ADHD"

Do some detective work looking for these common contributors:

- poor-quality sleep
- domestic unrest
- dietary insufficiency (see the following section)
- (in children) school bullying, boredom at school

---

3. SMARTEN YOUR NUTRITION

New insights suggest that ADHD may be triggered by *neuroinflammation*, which you learned about on pages 21 and 28. When researchers examined blood samples of more than two hundred people with ADHD, mainly adults, they discovered many had blood levels of antioxidants insufficient for their brain's levels of hyperactivity and hypermetabolism. That's why Dr. Mom's sage advice to "eat more fruits, vegetables, and seafood" (the three richest sources of antioxidants and anti-inflammatory nutrients) applies so well to people with ADHD.

In Dr. Bill's multi-decade experience as a pediatrician and an ADHD survivor and thriver, he's noticed that many children with this label are more sensitive to, and bothered by, unreal foods. Many times he's heard parents report, "He bounces off the wall when I let him have that highly sweetened cereal" or "His worst days at school are when he eats junk food." Dr. Bill's favorite: "She gets loopy on Fruit Loops."

Even more than adults, children with ADHD tend to be more sensitive to messed-up foods, such as modern gluten (see "Gluten Brain," page 326) and artificial chemicals. It seems that the "hyperactivity" of these children's head-brains may be triggered by *hypersensitivity* to artificial foods in their gut brains. This is why *eat pure* is Dr. Bill's main mantra in counseling these families. As Dr. Bill tells them: "Avoid GGG: gluten, GMO, and God-awful artificial food chemicals." He and his fellow pediatricians have long recognized that the greener children eat and live, the smarter they are. Science agrees.

Dr. Bill's experience in pediatric practice, Dr. Vince's experience in adult neurology, and their review of the scientific literature reveals that many people diagnosed with ADHD, at all ages, can be helped by the brain-smart nutritional approach we recommend in chapter two. And often, smartening one's eating—eating more "focus foods" and fewer fake foods—is the quickest way to quell quirky behavior. Make a food diary of what you (or your child) eats for a week. How many of those foods are on the *neurotoxin naughty list* (see page 75)? See what happens when you replace them with brain-smart alternatives.

We believe that much of ADHD begins with a "food disorder." The treatment? Just eat real foods.

Could ADHD begin in the gut? Over Dr. Bill's fifty years in pediatric practice, he has noticed that children who have the most learning difficulties also tend to have more intestinal difficulties, especially constipation. For this reason, often he will begin his consultation with parents by saying, "Let's start with giving your child good gut health." He then gives them a copy of his booklet, *Dr. Poo* (see AskDrSears.com/DrPoo), which explains how good gut foods make the microbiome (the second brain) smarter, which then makes the head brain smarter. Constipation is much more frequent in children diagnosed with ADHD, as is its opposite, chronic diarrhea. Not only are these kids' head brains out of balance, so are their gut brains.

Go clean and green. Some ADHD researchers believe the increase in learning and behavior problems in children over the past several decades parallels the rapid increase in the use of pesticides in our foods. Eat organic and from local farms as much as possible.

Go sugarless. One of the most common correlations parents of children with ADHD report is, "Once I cut sugar out of his diet, his behavior and concentration improved so much." The same seems to be true for adults with ADHD. Here, common sense and the science agree. Sugar spikes followed by sugar falls trigger the stress hormones epinephrine and norepinephrine, which in turn can trigger hyperactivity and poor concentration.

Going sugarless (avoiding "added sugar" in cereals and other packaged foods) is especially important at breakfast. Many people with ADHD eat the junkiest food at the time of the day when it contributes most to junk learning and junk behavior in school or at work. (The "sugars" that naturally occur in real foods, such as fruits and vegetables, are fine, mainly because these sugars are partnered with antioxidants, fiber, protein, and fat—nutrients that blunt sugar spikes.)

Go fish. Of all the nutritional changes that Dr. Bill has personally seen work in his medical practice at all ages, the most reliable—and the one that is supported by the most science—is to eat more seafood and/or take more omega-3 supplements. A 2013 study by Dr. Bill's friend and omega-3 mentor, Dr. Michael Crawford, showed that adolescent boys tended to have lower blood levels of the smartest omega-3s, EPA and DHA. Raising these levels is a simple change that is likely to have the quickest, greatest, and most lasting effect. (See the science and common sense behind this seafood suggestion in chapter two. And see what dosage to give for what your child has on page 45.)

Our thirteen-year-old daughter, Anna, labeled ADHD, didn't do well on drugs. They did help her focus, but they also sabotaged her sleep; she kept falling asleep at school. And they gave her tummy-aches. So, we got her off of them. After our consultation with you, Dr. Bill, where you tested

her blood level of omega-3s with the Vital Omega-3/6 HUFA Test™ and
prescribed more omegas, I'm happy to report: She's herself. She's able to
stay on task. The school thought she was still taking meds. Thank you!
−A grateful mother in our practice

MORE CHANGES YIELD FASTER RESULTS

While it's fashionable and sometimes works to do *one small change* at a time in adults, this slow-track approach seldom works with children. Children are growing fast and they need to feel fast changes. Make as many eating changes listed in chapters two and three as you can. Eating these smart foods in smart ways nearly always improves behavior and learning.

Food "medicine" before prescription medicine. Because ADHD drugs target symptoms, but not root causes, drug treatment alone rarely lasts. While many children show improved behavior and increased focus ability—often within a week of treatment—the child's brain eventually "outsmarts" the drugs, requiring a higher dose, a change of drugs, or the addition of new drugs. This is why in our own medical practices, in helping people with ADHD at all ages, we start with "nutritional psychiatry" and then, only when necessary, add pharmacological psychiatry. Whenever possible, skills before pills.

What's comforting about nutritional psychiatry is that there are no harmful effects, only healthful ones. Nearly every drug prescribed, especially for the very young and very old, comes with the baggage of unpleasant effects. So, because prescription drugs can show rapid relief, especially in people with debilitating ADHD, we sometimes prescribe both pills and skills, but gradually taper the dose of the pills as the skills start to click in.

NUTRITIONAL PSYCHIATRY FOR ADHD	
Neuroinflammation is one of the root causes of most ADHD.	Our healthy brain plan is high in anti-inflammatories.
Brains of people with ADHD may be more prone to oxidative stress.	Our healthy brain plan is high in the nutrients science most supports for ADHD, omega-3s and antioxidants.

COACH HAYDEN NOTES . . .

A few months ago, I ran into a friend who was worried about a possible ADHD diagnosis for her child. She only had a minute to talk and requested, "I know I need to go to the doctor and get an evaluation, and it's a long process, but I want something I can do NOW!" I told her to give her son only real food, cut out any foods with artificial colorings, sweeteners, and the like, and call me when she had more time. She called me a couple weeks later, thrilled because she had seen so much improvement after changing his diet. She never realized how much "fake color" he had been eating. She also commented on how his frequent stomachaches had lessened.

4. BEGIN THE DAY WITH A BRAINY BREAKFAST

In addition to cutting "added sugar" out at breakfast, preload the brain early in the day with the seven smart foods you learned in chapters two and three (see why and how, page 65). Eating a brainy breakfast prepares the brain to focus, instead of getting it hyped up and stressed by sugar spikes and falls. When you serve your brain junk food, you get back junk learning and junk behavior.

5. PREVENT "THE SITTING DISEASE" AND "THE INDOOR DISEASE"

Once they took recess out of school, Ritalin dosages went up. Any correlation? We think so! Often during an ADHD consultation, Dr. Bill asks the leading question, "How does your child spend their free time, say, after school?" "Playing video games" is the usual answer. Hearing this, Dr. Bill startles the parent by responding, "Your child has *MDD*—movement deficit disorder."

In the study of ADHD, there is a lot of controversy as to what it is, who has it, and how to treat it. However, there is one thing every ADHD specialist agrees on: the more hours per day someone with ADHD *moves their bodies* and *spends outdoors,* the lower the incidence and severity of the learning and behavior problems. These two "medicines" help everyone. (See related sections on "Why Movement Is Smart Medicine," page 96, and "Endnotes" for the Naperville Study, which showed how exercise improves behavior and learning.)

Imagine asking yourself: "Did I take my ADHD medicine today?" Translation: *Did I move more, sit less, and go outside and play?*

One of the Sears' children, who had, shall we say, a "focusing difficulty," did a lot of her spelling assignments while jumping on a mini-trampoline. Initially we thought, "That's quirky," then, "She's quirky." Movement mellows the mind; what she was doing made sense. That smart child, years later, is one of the coauthors of the book you are now reading, Hayden Sears.

DR. BILL'S FIRST TEACHER-THERAPIST

My fourth-grade teacher, Sister Mary Boniface ("Boniface" means "sweet face" in Latin) had me tagged. When she saw me fidgeting, not paying attention, and staring out the window (because I wanted to *be* out the window), she said with a sweet smile, "Billy, go outside and run around the schoolyard three times and come back and sit still." That was Dr. Bill's Ritalin. A high school teacher, Sister Mary Ursula ("Ursula" means "little bear"), prescribed the same therapy. Little did "sweet face" and "little bear" know that forty years later, outdoor exercise would top the list of best "medicines" for ADHD.

6. MATCH THE WORKING AND LEARNING ENVIRONMENT TO THE PERSON

When possible, craft classrooms and workplaces in a way that the brain will like.

Lobby for learning-friendly classrooms. Children deserve smarter classroom formats and teaching styles than the ones they get. They need large windows, natural lightning (no fluorescent lights), more space between desks, and windows that open to let in smart, fresh air. (We've noticed that newer schools, like newer hotels and office buildings, often have those horribly hermetically sealed windows that don't open.) Couple all this with a decrease in or lack of recess, where one of the only daily activities is walking to the cafeteria to get high doses of dumb foods, and you have a setup for not only ADHD, but a bunch of other "Ds."

Some very adaptable children can make this modern mismatch work, yet others can't, and those are the ones who get labeled. They are also the ones who get drugged—when we should be "drugging" our schools instead.

Our dream curriculum for students' mental wellness would include:

- a brainy breakfast
- movement before class
- mental breaks, or timeouts, for meditation
- more windows and natural light
- more time playing outside
- plants in the classroom, such as a Tower Garden® (see AskDrSears.com/tower-garden)

DR. BILL PRESCRIBES A "FOREST SCHOOL" FOR TOMMY

While working on this book, I was giving a lecture on brain health in St. Louis. As I was talking about how exercise helps you feel good, learn better, and make your own brain medicines—and actually moving my body to demonstrate—a five-year-old boy in the front row got up and started moving like I was. I had noticed that this young boy was really riveted by what I was saying and doing. Then I started talking about how movement makes more VEG (or VEGF, vascular endothelial growth factor, a medicine for the brain; see page 97). Tommy blurted out, "I want more VEG!" Naturally, his mom sat him down and whispered, "Tommy, that's not appropriate" (modern discipline-speak for "Don't do that!").

Seizing the opportunity, I invited Tommy to sit next to me on the stage. That smart decision upgraded the audience's attention, because Tommy would repeat, in kid language, my core concepts.

After the talk I congratulated Tommy's mother: "What a nice and smart boy you have! He is a genius. He's going to make a difference in the world." As a few tears came from her eyes, she informed me about Tommy's ADHD, his being one of the few kids to be kicked out of kindergarten, and how Tommy hadn't fit in at any school she had tried. She invited me to her nearby home, where she showed me his new "homeschool" classroom. I noticed there were a lot of windows in this room and it opened into a garden. "Tommy loves the outdoors and every day he helps me in the garden. He just seems to feel better outdoors."

"That's why they call it kinder*garden*," I concluded.

A few months later, she called to tell me that she found "Tommy's *forest school*." After expert counseling and much researching, she had found a school whose teaching methods better fit Tommy's brain. Much of the teaching and learning was done outside, weather permitting. What a perfect medicine match this mother found for her previously mis-fitting child!

Let the sunshine in. One of the top complaints school teachers have with kids tagged with ADHD is "He stares out the window." There are two reasons for this brain-healthy behavior. First, the natural colors of sunlight improve the brain's attention span. Second, kids stare out the window because they want to be out the window—their brains are saying "Go outside and play."

Neuroscientists have long taught that the brain works better with adequate exposure to sunlight. A 2010 study of 100 high schools in Michigan showed that the more green views (plants and trees) there were in the classrooms, cafeterias, and other places students hung out, the higher the grades. And a study of more than two thousand students showed that those who enjoyed the largest classroom windows learned better, especially in reading and math. Yes, sunlight is good medicine for the body and brain.

In schools, workplaces, homes, and everywhere humans live and think, the more windows, the sharper and mellower the mind.

Sunlight, nature, and ADHD. Researchers at the University of Illinois showed that when students did their activities outside, or at least while *viewing greenery*, ADHD behavior dialed down. This could be why teachers often observe that children show the worst behavior and difficulty learning in the wintertime, probably thanks to a combination of the indoor disease and the sitting disease.

Let the sunshine in.

BECOME A "STUDENT" IN YOUR CHILD'S CLASSROOM

If your child is struggling with ADHD symptoms, become a parent detective at the scene of the "crime." Ask your child's teacher if you can make a surprise site visit, and on some random day, quietly sneak into a back seat in the class and just be a fly on the wall. You may be surprised what you see that triggers disturbing behavior in your sensitive child. Here are a couple of observations Dr. Bill has heard from parents who have done this:

- "The kid in the seat behind her kept kicking her seat, and that bothered her attention."
- "He was getting bullied on the playground, but was too scared and ashamed to tell me."

After attending a few hours of morning instruction and observing playground activity (if there is any!), also take a stroll through the cafeteria to see what smart foods, or dumb foods, they are serving. You may walk away from this "office visit" with some of the real reasons why your child is behaving the way they are and become motivated to change the class to fit your child.

Design your brain-friendly workplace. To both brighten and focus your brain while working, follow our design prescription:

- Enjoy large windows that open.
- Look at beautiful multicolored plants.
- Use a desk that adjusts to sitting and standing (see Bob's story, page 257).
- Sit facing a window during meetings.
- Do isometrics when seated (see page 111).

8. TAKE MEDITATION BREAKS

"Hyperactivity" and hyperfocus can wear out both brain and body. That's why the brain-quieting perks of frequent meditation are especially helpful for people with ADHD.

Meditation benefits areas in the brain that promote focus and sustained attention, which is why taking mini-meditation breaks throughout the day could help teens and adults labeled with ADHD. Dr. Bill finds that periodic meditation breaks, especially swimming meditation (see page 153), are necessary during writing to help him relax and refocus.

Don't consider meditation as wasting time or doing nothing. On the contrary, meditation puts you in control of what's in your mind, or at least how to deal with it. In fact, modern science-based therapists teach meditation as a focusing and relaxing tool for people with all sorts of disorders. It has similar health benefits for the brain as going to the gym has for the body.

9. SLEEP WELL!

ADHD and poor sleep are partners in mental unwellness. Poor-quality sleep (especially when accompanied by poor-quality breathing, such as in obstructive sleep apnea) is one of the most overlooked triggers of ADHD. Children who sleep better also behave better, focus better, and learn better. If we line up the symptoms of people diagnosed with ADHD—unable to focus, highly changeable behaviors, and so on—they parallel those of sleep deprivation. And to add insult to sleep injury, some medicines that are prescribed as "focus pills" actually lessen the quality and quantity of sleep, and therefore interfere with the ability to focus!

Read chapter six on how sleeping smart preloads the brain to learn and behave, and recall Liram's story (page 263), where healing his dental misalignment and consequent obstructive sleep apnea also healed his ADHD.

10. ENJOY THE "CARRYOVER EFFECT"

This is one of the most important "treatments" we can "prescribe": discover your child's "special something." Every child is wired with at least one talent or hobby they enjoy doing—music, art, sports, acting, and so on. Then, once you've identified it, run with it. When the child's brain is happy, and they feel like they fit into this special talent, the overall boost in their self-esteem will carry over and then snowball to grow their social and learning skills.

Meet Adam. Dr. Bill had the privilege of being Adam's doctor since birth. Adam had trouble comfortably fitting into every class beginning in the first grade, but was fortunately gifted with wise parents who refused to "label" Adam and kept working to find ways to channel his unique brain wiring. During Adam's thirteen-year-old checkup, Dr. Bill asked about Adam's "special something." "Tennis!" both Adam and his parents shouted. (Tennis requires intense hyperfocus.) He became a star player on the high school tennis team and successfully competed in national events. Thanks to the carryover effect of his success at tennis, he not only managed most of his "Ds," but reframed them to work to his advantage. Dr. Bill recently hugged him off to the top college of his choice.

COACH HAYDEN NOTES . . .

I found that my son (a textbook bright, bored boy) was struggling with anxiety and feelings of sadness. His mind would race and his agitation would build to a point of being emotionally overwhelmed. I knew he needed a positive outlet for his unique brain. After experimenting with several activities, he fell in love with Rubik's Cubes. Something about the complexity of solving a puzzle mixed with the sensory dexterity for his fingers fascinated him and drew him in. It was something his brain enjoyed focusing on. In fact, the more he practiced cubing, the happier and more focused he was in other areas of his life at home and at school. And who knows—maybe one day he'll use this focus and dexterity as a surgeon! See a video of him in action and learn more at AskDrSears.com/ADHD.

"Wow, all this sounds like a lot of time and energy. Why not just use a focus pill?" you may wonder.

Dr. Bill has treated hundreds of children with ADHD and other "Ds" for nearly fifty years. In the process, he's learned that the ten steps you've just read work, and their effects last, because they fix the *root causes* of ADHD instead of just covering up the symptoms, as prescription drugs sometimes do.

Suppose, however, you've made all these smart changes and still see little improvement in focusing and behavior in yourself or your child. Here's your next step:

A VISIT TO DR. SMART, AN ADHD SPECIALIST

While you could go directly to a nearby specialist, such as a psychiatrist, psychotherapist, or university-based specialist in ADHD, it's often wise to start with a thorough visit to your primary care physician or pediatrician. Here are some time-tested tips on making the most out of your doctor's checkup.

What not to do. Don't try to sandwich in an ADHD appointment with an annual checkup. This happens to Dr. Bill all the time. Toward the end of a child's yearly checkup, just as he's closing the file, the parent opens up: "Oh, by the way, the school thinks Suzy has ADHD!"

Not smart! How can you expect to get life-changing advice in the minute left in your allotted appointment time? Remember, we are living in the age of volume-oriented medicine, where economic and management pressures force a doctor to see more patients per unit of time. Behavioral, psychological, and academic concerns require more time than most primary care doctors have, and receive less reimbursement from insurance companies.

Instead, here's what smart patients and parents would do:

1. Schedule a "consultation" or "long appointment" for "evaluation of possible ADHD."
2. Bring along your journal and a one- to two-page bulleted list, which will help you to quickly communicate to your doctor the concerns most in need of medical attention, such as:
   - "The school says he . . ."
   - "I see at home . . ."
   - "At work I . . ."
   - "With my family I . . ."
   - "These are the triggers . . ."
   - "These are the improvements we've tried . . ."
   - "These are the results we've seen . . ."
3. Plan to discuss your three top concerns, grading them on a scale of 1 to 10 in terms of severity. For example, "Our concerns, and

those of the teacher, about his learning and behavior at school are an 8. He can't focus on his math homework, and he just doesn't want to go to school." Or, "His work problem is a 9—he always comes home seeming so depressed."

4. Bring results of academic tests, or recent work evaluations.
5. Plan to review other treatments you have tried, such as neurofeedback.

Smart and experienced doctors and therapists try to read how concerned you are, how deeply you want or need advice, how conscientious you will be about following that advice, and how much the challenges you are having are affecting *quality of life*. The better you convey your concerns to your consultant, the smarter the advice you're likely to get.

Before rushing and reaching for a prescription pad, Dr. Smart will likely:

1. Do a thorough physical exam, including seeing if there is any airway obstruction that may compromise breathing and sleep. (See "Liram's story," page 263.)

2. Do a panel of blood tests for things that may affect behavior and learning, like omega-3 RBC index, serum iron levels, and vitamin D level (see blood tests for brains, page 319). In most visits, these tests usually come back normal, but as your wise doctor will tell you, "We want to be sure we are not missing a physical or metabolic cause somewhere in the body."

FOCUS TESTS FOR KIDS: YES? NO? SOMETIMES?

In my experience, the best "tests" for whether a child has ADHD are your journal and bulleted lists of the child's learning and behavior quirks. Smart parents are keen observers and accurate reporters, and they live with the child 24/7. While psychological and neuroimaging tests have certainly improved over the last decade, there are still problems that you should be aware of:

- Children are moody test takers. Catch them at a time of the day when they are well rested, well fed, and more alert (usually mid-morning), and the results are more reliable. Yet drag an unwilling, tired, and hungry child after school to take a test when he would rather be outside playing with his friends, and you have a setup for unreliable results. Because test results for ADHD in particular can vary with a child's mood, they could easily be different a few hours earlier or later.
- Interpretation of focus test results is particularly challenging for bright children who may become so bored at the perceived irrelevance of the questions on these tests that they tune out. They "fail" not because of an inability to focus, but because of the dullness of the questions. Give the same child a challenging test and they excel.

Dr. Bill remembers attending the first "session" with an ADHD specialist to have his daughter Lauren tested, as her school suggested. He'd politely demanded to be present during the test, promising to sit still in the corner and keep his mouth shut (which is hard for him to do), and he found the questions were so boring and irrelevant that it was hard for him to not only stay awake but not get angry. Then he noticed his daughter was just pushing answer buttons at random to get the test over with. At the end of the test, ten-year-old Lauren sincerely said to him, "Dad, you're wasting your money!"

Many test results are inconsistent because "inconsistency" is also one of the quirks of people with ADHD. One day their behavior and focusing ability is up, the next day it may be down. Some very bright children with ADD who are not necessarily hyperactive tend to have misleadingly "normal scores" because their motivation to please their parents is to hit the right buttons during the test.

A second visit with Dr. Smart. On your next visit, the doctor will want to review the records you brought and the results of the blood tests to see if there are any pieces of the puzzle that need to be put together before rushing into pills.

IS THE BEHAVIOR PROBLEM TRANS-SITUATIONAL?

Most of the time, but not always, if someone really does have a brain-pathway quirk that keeps them from behaving or learning to their full potential, the quirky behavior will occur in at least two, or all three, of these situations: at school/work, at home, and at play/in social settings. If it only occurs in one, then pills are unlikely to be necessary, since the person's "quirkiness" can be managed through skills, including changing the environment to prevent triggering the behavior.

Say your child behaves well at home, is fun to be around, and your parent's intuition doesn't have any alarms going off that something is wrong. Yet the teacher's reports don't match what you see. In that case, suspect the problem is a mismatch between the child and the school, and the smart "treatment" is to change teachers, change classes, or change schools. Or suppose the behavioral issues only happen at playtime. Suspect there is a bully lurking on the playground. Or if it only happens at work, suspect a poor job–brain fit.

SMART USE OF PILLS AND THERAPY FOR ADHD

As you know, the smart approach to management of ADHD is skills before pills, sometimes pills and skills, but never pills only. There are two reasons not to rush into pills, which by now may sound familiar:

1. Pills often help the symptoms—the ability to focus—but don't always treat the root cause. And with time, the brain sometimes outsmarts the medicine, at which point the harmful effects outweigh the helpful effects.

2. No brain drugs target just one "neurochemical imbalance." Even if we knew what player in the symphony orchestra is "out of tune" and causing the inability to focus (which is difficult if not impossible to determine), all the players are interconnected, and what drugs one player, drugs others. Suppose a drum is playing too loud, and you are prescribed a drug to dampen its sound. The drum may dial down, but so may the strings. And the trumpet may dial up! Many "focus pills," as we call them, help you pay attention during school or work hours, yet dampen appetite at mealtimes and lessen quality sleep at night.

When it comes to giving medication to children, there's a third good reason to avoid pills if you can: the goal of parenting and education is to teach your child tools to succeed in life. Instilling a *pill dependency* at an age when a child's brain is most permanently impressionable is risky. *Skill dependency* rather than pill dependency is a smarter tool for all ages.

Still, after discussing all your options, you and your doctor may decide to add medication, or pills, to your skills.

Many ADHD researchers believe that it is caused by a neurochemical imbalance. This makes sense, since many people with ADHD also struggle with other "Ds," such as OCD, BPD, depression, and generalized anxiety disorder. Some studies reveal that the neurotransmitter GABA, which calms impulsivity, is lower in some children and adolescents. ADHD researchers also suspect that the reason stimulant medications—or "focus pills"—such as methylphenidate (brand name Ritalin) and its newer cousins work is that they correct an imbalance in the alertness and focus neurotransmitters. They bring the level of the focus neurohormones to just the right level for the individual, not too low and not too high.

This effectiveness comes with an uncomfortable side effect. As with gravity, what goes up must come down. When taking a focus pill during the day, some ADHDers report, "It worked like a charm!" Yet, as with caffeine, when the pill wears off, they may experience an evening crash. Or, if the "focus pill" is taken too late in the day, because it is an "arousal drug," people report, "I was so wired I couldn't sleep." Since sleeplessness is one of the most prominent yet often overlooked triggers of ADHD, the last thing you want is to sabotage your sleep. The euphoria and elation this medication can offer during the day often leaves its user both jittery and sad in the evening as the pill's effect wanes.

An additional caution: Stimulants may also interfere with other medications adults with ADHD may already be taking, such as those that manage blood pressure.

If you and your doctor do nonetheless decide medication will be helpful, here are some things we recommend you keep in mind:

Emphasize the skills. Continue all the things that we know help balance the brain's behavior: eat smart, move smart, think smart, and sleep smart. Remember, focus on the skills first, and then add help from the pills.

Theoretically, the pills increase the levels of neurotransmitters in the brain, such as serotonin, GABA, dopamine, and acetylcholine. But skills make the membranes of the brain cells smarter, the nerve transmission smarter, and the neurotransmitters smarter—increasing the effectiveness of pills by improving how well the natural neurotransmitters that pills stimulate work.

Skills are safer than pills. When it comes to any treatment option, you want to make sure you're considering the risk/benefit ratio. Translation: You want medications and therapies with the highest possible benefit and the lowest possible risks. Skills carry zero risk. Pills are a little more complicated.

One of the benefits of a stimulant medication, such as methylphenidate, is that when it really works, you will notice obvious improvement in the problem behaviors within a few days to a week. (In our medical practices, when we ask a person, "Is the medication helping?" and they waffle a bit, such as, "Well, I think it is . . . ," it probably means the medication isn't having a real pharmacological effect.)

The most common undesirable effects (called adverse reactions) of stimulant medications are decreased appetite, sleep problems, and, in children, long-term growth suppression. Studies have shown that children medicated with methylphenidate tend to have an average growth slowdown of almost an inch in height and six pounds in weight over a period of three years. Other adverse effects include an increase in blood pressure, heart rate abnormalities, and manic outbursts, though these are much less frequent.

Explain the pills. Your doctor may begin by explaining how to use them. "Tommy, I'm going to give you some pills. You take one every morning on school days." (If the ADHD is trans-situational, as discussed on page 279, your child may need to take the pill every day, rather than doing "drug holidays" or stopping the medication during the summertime.)

Then, they may invite the child to observe the effects as neutrally as possible. "I want you to tell us how they make you feel." (If you plant "These pills are going to help you focus better" in a child's already sensitive and creative mind, you may trigger the placebo effect [see page 283].) To avoid keeping them on a medication that isn't making a difference and risking the side effects, better to let the child more objectively report, "They're helping," "They're hurting," or "No change.")

It can also be useful to outline the plan. "We're also going to use the smart pill method of *start low and go slow*. Because your situation is not that serious, we're going to start you with a lower dose and gradually increase it, if necessary."

Finally, they should establish an endpoint. "Over time, say six to twelve months, we'll plan to slowly wean you off the pills as your smart skills take over. We want you to be happy at school, with your friends, and at home, and to love to learn."

IS IT THE PILLS OR YOUR WISH THAT WORKS?

We've talked about the placebo effect in terms of anxiety and depression (see page 131). But the way the brain works is a setup for pill dependency, which is why the placebo effect is sometimes called "the belief effect." The more your child believes the pill will work, and the more you believe the pill will work, the more it's likely to. And, as a parent myself, I understand because you love your child so deeply, you really just want him to learn, feel, and behave better, and are less concerned with how the pill works. Studies reveal that the placebo effect is likely to be higher (around 30 percent) in children with ADHD than in adults with ADHD. Makes sense, since children are not always the most accurate reporters, and loving parents want their child to focus better.

Start with one drug only. Most ADHD specialists recommend beginning with methylphenidate (Ritalin and its newer cousins) only, and not in combination with an amphetamine—Adderall and others. Why *amp* up an already hyped-up brain?

In addition, easy availability of these drugs has led to a dangerous rise in dependency on and abuse of ADHD drugs containing amphetamines, including among teens *without* ADHD. Parents, be sure to monitor your teen's exposure to "amphetamine parties." Watch for red flags, such as coming home obviously high on something.

Journal, journal, journal. It's very important, when taking medication for ADHD, to keep a journal. When you don't, it's easy to end up on a drug or dosage that is not the right fit for you. If it's your child who is on the pills, you can feel so desperate to see your child feel and do better that you want to imagine the pills are working.

Sit down every day, or every other day, just for a few minutes, and record how the pills are making you feel. Be as honest as you can. There are no right or wrong answers. Your answers will help your doctor prescribe the right pill at the right dose.

NEUROFEEDBACK IN A NUTSHELL

Whether you decide to use medication or not, your doctor may also refer you to a specialist in neurofeedback. Think of neurofeedback as weight training for the brain; you exercise the brain pathways that help you concentrate, the way learning gymnastics or piano lessons grows neuro-pathways that focus on movement, balance, and music. Neurofeedback training is kind of like playing a video game, making it particularly appealing for children. You control what you see and hear on the screen by controlling your level of concentration.

Neurofeedback is a wonderful workout for the wandering mind. When your mind wanders, the video game stops and prompts you to regain your focus. In other words, the screen "feeds back" to you what nerve pathway changes it wants you to make; you do it, and your attention pathways grow.

The way it works is that sensors are comfortably placed on your scalp to record brainwaves that reflect changes in thoughts and attention. This not only lets you know when your attention has wandered, but allows a trained neuroscientist to evaluate the effects of the neurofeedback on the actual functioning of your brain.

Our friend of nearly fifty years, Lynda Thompson, PhD, director of the ADD Centre, coauthor with Dr. Bill and Martha of *The A.D.D.* Book, and coauthor of the textbook, *The Neurofeedback Book*, explains:

> To help a person improve their attention, you monitor their brainwaves (the EEG) and give visual and auditory feedback through a game, animation, or a DVD. The feedback plays when the brainwave pattern shows calm focus and stops if the person drifts off. With enough training, the brainwave pattern shifts: less power in the tuning out waves (theta) coupled with higher amplitudes of calm waves (sensorimotor rhythm) and thinking waves (beta) [meaning the distraction/clutter of the brain dials down, and the "calm" centers dial up]. Focusing pathways grow.

Research done in Montreal showed that children with ADHD who did forty sessions of biofeedback training not only managed their ADHD symptoms without the use of medication but additionally showed changes in brain activation as measured by fMRI. This included increased activation in the substantia nigra, where dopamine is produced. More recently, research from the Universite de Montreal indicated increases in both gray and white matter volume after twenty sessions of neuro-feedback training done with healthy university students. Brain exercise actually bulks up the brain.

For twenty-five years, the ADD Centre has offered this training in Mississauga and Toronto, treating both children and adults who want to improve self-regulation of attention and self-regulation of emotions. We start by discussing goals and doing an assessment that includes single-channel EEG to check the individual's brainwave pattern. Then we set a program to optimize brain function and the client does their brain exercise twice a week. He/she gradually shifts their brainwave patterns toward more desirable ones, such as less busy brain (high frequency beta activity) and more calm waves in someone who ruminates and has trouble falling asleep.

A fifty-six-year-old lawyer, one of our first adult clients, reported that the neurofeedback training took two hours off his working day because of improved focus and efficiency. He then sent his secretary for training!

Thank you, Dr. Lynda!

REFRAMING THERAPY FOR KIDS

Because children with ADHD (and other "Ds") are often super-sensitive and supersmart, when you tell them you are taking them to a therapist, they often conclude, *Something is wrong with me. My friends all get to play soccer after school but I have to go sit in this dumb, dark room instead.*

It can help to program your child with something like, "Think of your therapist like a *coach* who helps you with learning skills. Just like you listen to your sport's coach and do what he says, you have to listen carefully to your therapist and do what she says."

ADHD UPDATES

The understanding of the root causes of ADHD continues to improve, and newer and safer treatments are being developed all the time. For updates, see AskDrSears.com/ADHD.

CHAPTER 10

≈≈≈≈≈≈

DRINK SMART: PREVENTING ALCOHOL DISEASE

I f doctors voted on what "drug" damages the most brains, the winner—really the "loser"—would be alcohol.

Within our own families of friends, alcoholism is the number-one cause of unhappiness. Some have lost their marriages, their money, and their minds due to the brain-damaging effects of excess alcohol. The good news is, some of them have addressed and overcome their excess alcohol use, and became wiser and healthier people. Others, sadly, have let alcohol control them instead of them controlling alcohol, and entered a downward spiral of deteriorating mental and physical health.

We include this important chapter on alcohol because it is often overlooked as a contributor to mental illness. Because of the effects of excess alcohol on the body and brain, especially occurring at younger ages, doctors now address this problem as *alcohol disease*.

While not a popular label among drinkers, *alcohol disease* makes sense. You eat too much added sugar and other highly processed foods and you get the disease of diabetes. To treat or prevent diabetes, you begin by eating less sugar. If you drink too much alcohol, you get alcohol disease. And to treat or prevent it, you begin by drinking less alcohol—and drinking smarter when you do.

Drinking smart is especially important for brain health because, as the famous GRANT Study (see "Endnotes") revealed, alcohol disease is one of the top contributors to developing Alzheimer's disease.

HOW SMART DO YOU DRINK?

To figure out how best to use this chapter, ask yourself: Where do you personally fit in the spectrum of drinking?

Do you drink to *enjoy the taste* and how alcohol, especially wine, "pairs" with (enhances the taste-pleasure of) certain foods? Two or three evenings a week, you drink a glass of wine with meat, pasta, or whatever pairing you like. You only pair alcohol with meals and never drink on an empty stomach. But on other evenings you don't drink, and you can easily go without alcohol for a week.

If this is you, no problem; enjoy! You can read quickly through this chapter for tips on how to remain where you are.

Do you have an *alcohol dependency*? Do you drink because you need an *-ant*: a relax*ant*, antidepress*ant*, or escap*ant*? A drink after work helps you relax, but you know when to say when.

If this is you, read on for how to keep from crossing the line between occasionally drinking just to relax and "must drink to relax," or *alcohol dependency*, which then can deepen to *alcohol addiction*, and finally to *alcohol disease*.

YOUR BRAIN ON ALCOHOL

Alcohol is one of the most harmful drugs you can buy without a prescription. Yes, alcohol is a drug! Like all drugs, it has harmful side effects. And like many drugs, the higher the dose and the longer it's used, the greater the harmful effects. Yet, unlike many pharmacological drugs, which are not addicting and not difficult to wean off of, alcohol is addicting. The more you drink, the more you want, and the more you have, the worse its effects.

If that isn't enough to prompt you to pass on that refill, consider the following:

Alcohol is a high-calorie sugar. We know from chapter two that excess sugar is what makes us fat. Double that for the sugar in alcohol. The fermentation process that changes grapes into wine, for example, nearly doubles the calories in each gram of alcohol. Obviously, alcohol is not a diet drink.

As you learned on page 21, "Avoid spikes" is one of the most important pieces of advice for brain health. The faster a carb spikes, the higher your blood sugar, and the higher your blood sugar, the sicker your brain (and body) gets. Suppose you lined up "contestants" for the Dumb and Dangerous Carb Award—sweetened beverages, white bread, marshmallows, doughnuts, and alcoholic drinks. Alcohol wins! Or, more correctly, loses. Alcohol, especially hard liquor, is more rapidly absorbed in the upper intestine than are other carbs.

Fast track into blood- stream

Biochemically, alcohol can be described as the molecular garbage dumped by yeast cells burning sugar, or "yeast poop." This mischievous "yeast poop" manages to worm its devilish way through the protective lining of the gut and into the blood. Alcohol has a peculiar quirk that accounts for both its woozy effect, which some like, and its brain-damaging effects, which aren't smart. It mixes easily with both fat and water, meaning it quickly gets through the gut's protective lining and into the blood. And which organ soaks up all this alcohol-saturated blood? The brain, because it is one of the highest blood-users of any organ. The damage starts there.

Alcohol dumbs down smart fats. A tissue is only as healthy as the cells composing it, and a cell is only as healthy as the cell membrane that protects it by keeping out neurotoxins that try to get through it.

Because brain cell membranes are mostly fat and your brain is a "fathead," alcohol's fat-damaging effects particularly bother your brain. An interesting study from the National Institute of Alcoholism and Alcohol Abuse at the National Institutes of Health showed that, at least in cats, excessive alcohol decreases the concentration of DHA (the smartest fat) in brain and eye tissue. One take-home message of this study (done at one of the nation's most trustworthy research centers) is that people who already have low dietary and blood levels of DHA—that is, those not eating an intentionally brain-smart diet—should definitely not drink alcohol.

Alcohol is both a relaxant and a depressant. Alcohol, in low amounts, perhaps one drink, may calm anxiety and behave like a temporary tranquilizer, which is why some people start to drink to self-medicate.

Alcohol may increase electrical activity in the same brain systems affected by some pharmacological drugs. Low doses (again, perhaps one drink) may lead to a feeling of pleasure and euphoria. It works on similar brain circuits as "benzos," the class of tranquilizers that includes Valium (alcohol is often dubbed "liquid Valium"). In some ways it also resembles opium, because alcohol releases morphine-like compounds, called endorphins, that trigger the brain's pleasure circuits. Yet, as you will soon learn, alcohol abuse can worsen depression. Some people overdrink to escape from the debilitating feelings of depression, but are unaware that alcohol can in the process deepen depression.

Another very important point is that the "alcohol effect" is different at different doses, and in different brains. While some people can drink smart, limiting their amount and frequency, others suffer the *creep effect* of slipping into an alcohol dependency. (See "Beware of the Creep Effect," page 302.)

One of our favorite quotes on alcohol is from neurobiologist Steven Braun, author of *Buzz: The Science and Lure of Alcohol and Caffeine*. "Alcohol," he explains, "is a pharmacy in a bottle: a stimulant/mood-altering drug that leaves practically no circuit or system of the brain untouched. Unlike drugs, such as cocaine and LSD, which work on only one or a handful of brain circuits, alcohol is like a pharmacological hand grenade . . . In other words, unlike many prescribed drugs, alcohol harms many circuits of the brain."

IS ALCOHOL A "HEALTH DRINK"? NO!

You may have read that alcohol, at least in small doses, has healthful effects, such as lowering odds of developing heart disease. But it's important to know that not all the experts believe these conclusions. They are often the result of statistical manipulation and/or suffer from the *correlation, not causation* fallacy, meaning that even if statistics show that people drinking a certain amount of alcohol suffer less from a certain disease, that doesn't mean the alcohol had anything to do with the outcome. The correlation could be meaningless. We believe it's the seafood, olive oil, and real food in the Mediterranean diet that makes it healthy, not the wine. Yes, *resveratrol*, the healthy antioxidant in red grapes, is present in wine, but not enough for wine to be considered a "health drink." You're better off just eating grapes! The general consensus among the most trusted health experts is: *If you don't already drink, there's seldom a medical reason to start.*

Alcohol blocks neurotransmitters. Neurotransmitters are, as we learned in chapter one, the molecular language of the brain—the way that brain cells talk to each other. Change your biochemical language, and you change your brain talk. That's what alcohol does.

One of these "talk chemicals" is glutamate, also known as the "excite" neurotransmitter. Alcohol reduces glutamate firing and thereby helps the brain dampen the anxious centers of the brain—just what a person suffering from manic or anxious behavior would want. Although this short-term effect sounds appealing, it has serious long-term ramifications.

A balanced brain has just the right amount of glutamate neurotransmitters (which excite the brain) and GABA neurotransmitters (which calm the brain down). But years of heavy drinking can disturb this balance and, as you already know, in a brain out of biochemical balance, the neurotransmitters dial up too high (anxiety) or too low (depression).

In some people with early alcohol disease, the brief initial "high" of drinking, which lasts 20 to 30 minutes and stems from a *dopamine dial-up*, is shortly followed by a *dopamine drop*, where levels fall below where they were before the person started drinking. These people drink to get undepressed—only to wind up later in a deeper depression than the one they were in before.

> Chronic alcohol use changes the brain's reward system by, perversely, dampening the pleasurable effects of alcohol and increasing the alcoholic's craving for alcohol. The alcoholic's intense craving is in part a result of the permanent increase in "alcohol-sensitive" buds on the brain's neurons—a potentially life-long change in brain structure and function.
> —Anderson Spickard Jr., MD, author of *Dying for a Drink*

DOUBLY DEPRESSING

One of the smartest bits of wisdom when it comes to "the Ds" is: "Don't drink if you're depressed." The problem is, lack of inhibition is often a side effect of depression, and depressed people often resort to alcohol as an escape. Because depression can be so debilitating, people prone to depression, such as those with bipolar disorder (BPD) or other mood imbalances, are at increased risk of alcohol disease because they are used to self-medicating, with alcohol as their "medicine." As alcohol use deepens their depression over time, they tend to increase their drinking accordingly, which only serves to damage their brain and body more.

HIGH-FUNCTIONING HEAVY DRINKERS

Some people seem, on the surface, to have a higher tolerance for alcohol; they are able to get drunk at night and still perform their job well the day after. While this higher tolerance may seem helpful, it often backfires. Being able to hold down a job and a relationship requires a person to continue to deny that they have a "disease," which prevents them from getting help. By the time this high-functioning person finally starts feeling alcohol's effects on their job and family, brain and liver damage already may have occurred. The paradox is that high-functioning people with this disease can have a higher risk of brain and body disease from alcohol simply because they do not seek help early enough.

OTHER WAYS A BINGE BOTHERS YOUR BRAIN

Let's follow a binge from bottle to brain, and into other organs, to understand how appropriate the term "wasted" really is.

Alcohol puts you physically off balance. Alcohol dials down the *cerebellum*, the center in your brain that promotes body balance. This is why, after you down too many drinks, you are more likely to fall down.

If, when you fall, you happen to hit your head, it can shake up your brain, causing a concussion, or what is called traumatic brain injury (TBI). TBI causes *neuroinflammation*, which adds to the brain-damaging effects of alcohol.

Alcohol is a central nervous system depressant, which means it slows your reaction time. That's part of why driving under the influence is so dangerous. Your physical responses—for example, quickly hitting the brakes or swerving when needed—are delayed.

Sometimes a fall and consequent TBI are so severe that you suffer a spinal cord injury. You may suffer paralysis, either quadriplegia (all limbs paralyzed) or paraplegia (two limbs paralyzed), and need to rely on a wheelchair for mobility for the rest of your life—all because of a one-night binge or, in this case, a one-night fall.

Alcohol damages your ability to think rationally. Not only is your body out of balance when you're drunk—so is your brain. We've already seen how alcohol can affect neurotransmitter balance. Alcohol also damages your *frontal cortex*—the rational or "make wise choices" center of the brain. This allows the more impulsive parts of your brain to take over, prompting you to take more risks—like continuing to drink.

The frontal cortex is also responsible for governing your *inhibitions*. When it's impaired by alcohol, you are vulnerable to doing things, or letting others do things to you, that you would otherwise have the wisdom not to—for example, having sex, committing a violent crime, or driving while drunk.

A person "under the influence" of alcohol is no longer under the influence of the rational part of their brain. Their logic center has been put on mute, and they can't hear that part of their brain telling them, "You're too smart to do this," "You deserve better," "Don't take this dumb risk that may cost you your marriage or your job." This doesn't excuse what they do, but it does often explain it.

Doctors label alcohol's effect on the brain as "stupefying." Under the influence you tend to say and do stupid things.

Alcohol fires up the brain's addiction center. Besides weakening the inhibition center, alcohol also dials up your addiction center so it's harder to make healthier choices, especially in the areas of food and sex. You're more likely to overeat and make decisions you'll regret, including ones you may have difficulty completely erasing from your memory—a setup for post-traumatic stress disorder (PTSD). With your crave-control center impaired, it's hard *not* to drink more. And the more you drink, the more you impair your brain, to the point where a friend or loved one might rightly conclude, "They cannot have just one drink; they cannot not drink."

"BUT I DON'T REMEMBER . . ."

One of drinking's cruelest effects is on memory. When we drink, especially in excess, we tend to forget information we recently learned.

In studies comparing college students who abstained from alcohol with those who binged at a party after a few days of learning, the drinkers were more likely to forget what they learned—they forgot up to *half as much* as the smart students who abstained. (Notably, trying to "catch up" with a few good nights of sleep after did not retrieve the memories or heal the forgetfulness these students experienced immediately after their binge.)

The extreme version of this effect is memory blackout. Because of these blackouts, people with alcohol disease often have little memory of the trauma drinking causes themselves and others. As a result, they can't learn from their mistakes nor vow not to repeat them, instead fooling themselves into thinking they don't have a problem.

By now, three centers of the drunk brain are temporarily damaged: the body-balance center; the rational, or think-smart, center; and the crave-control center. By this point the drinker has, sorry to say, lost part of their mind, either temporarily or permanently—depending on whether they dial down their AD (the sooner they get and stay sober, the sooner they can heal their broken brain). But the brain is hardly the only part of the body damaged by alcohol.

UNSOUND MIND, UNSOUND BODY

When the brain gets sick, the body feels it, too. Because of the neurochemical, or text-messaging, connections between the brain and all vital organs, the drinker's unhealthful habits damage other vital organs of the body, which, consequently, further damage the brain through increased inflammation.

Booze belly. Carb spikes from alcohol, like carb spikes from food, lead to belly fat. When these belly fat cells get too fat, they dump inflammatory sticky stuff into the bloodstream, which then collects in the brain, worsening neuroinflammation. This is why drinkers suffer early Alzheimer's disease; as mentioned in chapter eight, Alzheimer's is thought to be caused by an excess accumulation of sticky stuff (in this case, beta amyloid and tau protein) that damages neural wiring and firing.

Lousy liver. As we've seen, one of the main reasons alcohol affects the brain so heavily is that the brain is richly vascular. If a lot of alcohol gets into the blood, a lot gets into the brain. As your blood alcohol level rises, your lovely liver, the body's detox center, comes to the rescue, working overtime to break down the alcohol before it breaks down the brain. Yet drinking too much for too long eventually overwhelms the liver's waste management system. The not-so-lovely result: *cirrhosis* (scarred and dead liver tissue) and *fatty liver disease*. Without your liver, you don't live very long.

DR. BILL'S EARLY LESSON ON DUMB DRINKING

As a young teen working maintenance jobs in the homes of drinkers, one of the first causes of death I learned about when I asked "Why did our friend die?" was cirrhosis of the liver. The term "cirrhosis" was such a scary term that my vulnerable teen brain was programmed—for life—that whatever that was, I never wanted to get it. And that correlation between overdrinking and early death stuck.

Painful guts. Your stomach has two lines of defense against the toxins you ingest: high levels of acids, which destroy foreign chemicals, and a thick mucus lining, which prevents the toxins from passing through into the bloodstream. Unlike many foreign chemicals, alcohol molecules resist the acid attack, and their biochemical ability to dissolve in both water and fat allows them to penetrate the mucus shield.

Because the otherwise protective lining of the stomach is less protective against alcohol, heartburn, stomach ulcers, and many other tummy troubles increase while drinking. Alcohol stimulates the acid glands of the stomach to produce more acid, further weakening the already damaged stomach lining. As a result, regular drinkers usually suffer from gastroesophageal reflux disease (GERD).

Along with overproduction of acid, alcohol has a double digestive fault: it also decreases secretion of pancreatic digestive enzymes, leading to indigestion. If undigested alcohol gets all the way down into the lower gut, it can also kill friendly bacteria in your microbiome, your gut-health pharmacy (see page 85). It contributes to leaky gut syndrome, as well: alcohol damages the gut lining just as it does the stomach lining, which then allows stuff to get through that shouldn't.

SLEEPLESS ON ALCOHOL

Will a drink or two after work affect your sleep? Yes, and for the worse! Alcohol is not a sleep aid, it's a sleep disruptor.

First, alcohol disrupts sleep with frequent awakenings. Besides causing sleep-unfriendly gut pains, alcohol blocks the release of anti-diuretic hormones, which naturally perk up at night in order to lessen your trips to the toilet. Consequently, drinkers have to pee more at night. (This effect may also cross over into the daytime hours, leading to dehydration and sometimes even low blood pressure.) Even though the next morning you may not remember how often you woke, you're likely to feel "unrefreshed."

Second, when you're drunk, you don't dream. Alcohol steals REM sleep, and as you learned in chapter six, this dream sleep is key to our emotional well-being.

Contrary to popular belief, "sleeping it off" is not really neurologically correct. As the science shows, the sleep that follows a night of drinking is neither normal nor healthy.

I'm sleep deprived.

THE MORNING AFTER

One of the biggest complaints that drinkers have about drinking is the hangover they experience the next morning. They have a headache due to alcohol's effect on the blood vessels in the brain: when "drunk," the vessels dilate and constrict, producing brain pains. Throughout the next day, they may also experience hypersensitivity to light and sound. Their stomach is queasy and rumbling due to increased stomach acid. As the liver diverts all of its efforts to detox the blood of alcohol, it has less machinery to devote to clearing the blood of other metabolic waste, such as lactic acid. The resulting buildup of lactic acid contributes to feeling "lousy" and "not myself."

IS OVERDRINKING WORTH IT?

These scary statistics may prompt you to stop and think before you overdrink. Drinkers suffer a higher risk of the following ailments:

- Pains in the gut (heartburn, irritable bowel, etc.)
- Heart disease (alcohol allows too much cholesterol to be pumped into cell walls, hardening them)
- Diabetes
- Depression
- Early Alzheimer's
- Osteoporosis (excessive alcohol drinking increases acidity, which contributes to bone loss)
- Broken marriages
- Abusing their spouse and children
- Losing their eyesight
- Losing their libido
- Losing their life

More than one million people per year are admitted to hospitals due to alcohol-related injuries and illnesses—another reason we include "drink smart" in our healthy brain plan.

SEVEN RULES OF SMART DRINKING

While alcohol-free living is best for your brain, if you still want to drink, try these tools for smarter drinking:

1. REMIND YOURSELF THAT DRINKING CAN BE DUMB AND DANGEROUS

Thoroughly read this whole section on alcohol and try to prime the permanent file cabinet in the rational center of your brain to read: DON'T DRINK! or DRINKING IS DUMB! Often, if this doesn't eliminate your desire to drink, it will at least decrease it. Visualize the illustration on page 300 when tempted to binge.

2. SIP SLOWLY WITH MEALS

Drinking on an empty stomach wins the "dumbest drinking" award. So-called happy hour, where a person downs a cocktail before dinner, is the worst time to drink.

Eat more, absorb slower. As you'll recall, one of the unhealthy quirks of alcohol is its rapid absorption into the bloodstream, where it spikes blood sugar too high, too fast. Food slows this absorption. Pair slow sipping with high-fiber and high-fat foods, such as salads with olive oil and avocado, to slow down the absorption of alcohol.

Eat more, absorb less. Alcohol is "detoxified" in the stomach and liver by the enzyme alcohol dehydrogenase. The longer alcohol remains in the stomach, the more it is detoxified before it enters the small intestine, where it gets even more rapidly absorbed. Food holds alcohol in the stomach longer, giving your body's natural detox system more time to work.

3. CHOOSE THE LEAST UNHEALTHY ALCOHOL

What you choose to drink can make a big difference in the potential for health-harming effects.

- Wine is usually easier on the gut and brain than hard liquor, because of its much lower alcohol content. Remember, *hard liquor is hard on the brain*—both your gut brain and your head brain.
- Beer can be an unhealthier choice than both wine and hard liquor because it is higher in carbs, which leads to a beer belly. Fruity cocktails, too, such as piña coladas, are also likely to put on the belly fat.

DIFFERENT SHOTS FOR DIFFERENT FOLKS

Because how fast the gut breaks down alcohol and how quickly it gets into the bloodstream can vary widely in different people, know your tolerance. If one drink bothers you, your body is telling you to drink smaller amounts more slowly. If you are genetically a "fast flusher," you may be able to tolerate more alcohol with less harmful effects. If you are a "slow flusher," meaning you don't have the biochemical enzymes to break down alcohol as quickly, be a teetotaler or drink smaller amounts more slowly.

4. BEWARE OF THE CREEP EFFECT

This effect sees you justifying your drinking by saying that you'll limit yourself to just one glass of wine at your evening meal, but then you notice yourself saying, "Well, maybe I'll just have a glass and a quarter . . ." then "a glass and a half . . ." To curb the temptation to "have one more glass," pour your "safe amount," say one 5-ounce glass, then put the bottle away.

HOW MUCH ALCOHOL IS SAFE TO DRINK?

This varies from person to person according to their individual metabolism (and gender; see page 305), and depends on three other key factors:

1. *The concentration of the alcohol.* "Hard liquor" can be as much as 85 percent alcohol; wine is between 12 and 14 percent.
2. *How fast you drink it.* Slow sipping slows absorption into your bloodstream.
3. *With what food you pair it.* Healthy fatty foods like cheese, guacamole, or bread dipped in olive oil help to slow alcohol absorption.

In general, researchers who study alcohol recommend keeping your alcohol intake to no *more than one drink a night*, where "one drink" means either a 12-ounce bottle of beer, a 5-ounce glass of wine, or a 1.5-ounce shot of 80-proof hard liquor (all have the same alcohol content). (See "Endnotes" for research on how much alcohol is safe to drink.)

12 oz = 5 oz = 1 1/2 oz

what is "one drink"?

5. TAKE BREAKS

Take an occasional multiweek vacation from drinking. You'll miss it less than you think. If you do find that you miss it horribly, it might indicate you have a dependency or addiction, and you may want to consider a trip to your friendly neighborhood Alcoholics Anonymous meeting.

KEEP ALCOHOL OUT OF SIGHT

Out of sight, out of bloodstream and brain. Keeping wine and liquor bottles inside cabinets lessens the temptation to have "just one more."

PARENTAL GUIDANCE RECOMMENDED

Parents, we have a dangerous problem. One in two high school seniors drinks, and one in three reports being drunk in the past few months. And among college students, 1 in 2 young men and 2 in 5 young women (approximately 1 in 4 students overall) are *binge* drinkers, meaning that most weekends they consume four or more drinks over the space of a couple hours.

Why is this such a big deal? Teen brains are particularly vulnerable to the effects of alcohol and addiction, and adolescents who binge-drink may permanently damage their brain. Early drinking is also a risk factor for alcoholism. Two out of five adolescents who begin drinking by age fourteen will become addicted to alcohol. Binge drinkers in particular are at very high risk for doing stupid and dangerous things, and for getting into fatal car accidents. In fact, binge drinking is the number-one health problem on most college campuses.

To prevent your own child from becoming a sad statistic, raise a teen teetotaler. Model how to "drink smart" at home and during family vacations. But keep in mind that parents who have a family history of addictive drinking may overreact and have a rigid attitude toward any drinking, which can backfire: studies show this may actually increase a child's desire to drink. College students are often tempted to experiment with their home "forbidden list" once they are on their own.

6. AS YOU AGE MORE, DRINK LESS

Our alcohol garbage-disposal system dials down as we get older and our gut becomes more sensitive to alcohol's effects. Also, the older we get, the more pills we take that compete with the body's alcohol-detox machinery. Overconsumption of alcohol also causes a person to produce more estrogen, which is why men with alcohol disease often grow "man boobs."

After age fifty, most men and women should drink less. As a general guide, seniors—especially those over sixty-five—should cut their drinking in half.

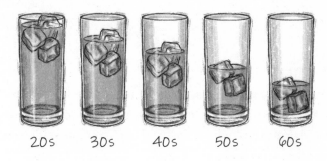

20s   30s   40s   50s   60s

ALCOHOL IS NOT FEMALE-FRIENDLY

Women are likely to suffer the harmful effects of alcohol more quickly and at lower doses than their male counterparts. Here's the biochemical reason: women have less of the enzyme responsible for breaking down alcohol, alcohol dehydrogenase, so more of what goes into the mouth gets into the bloodstream. Men should take note, however, that they "age out" of this protective edge after fifty, when their production of alcohol dehydrogenase decreases.

WHO SHOULD NOT DRINK?

Because of changes in how you metabolize alcohol, do not drink if you:

- are pregnant (or trying to get pregnant).
- have cholesterol problems.
- suffer from depression and/or anxiety.
- are prone to anger and irrational thinking and behavior.
- already have Alzheimer's or mild cognitive impairment.
- already have domestic problems.

7. KNOW WHEN TO SEEK HELP

Like other diseases, alcohol disease is best cured by prevention. The key is to prevent tissue damage before you feel it or blood tests show it. You don't want to wait until your liver prompts you: "I'm being wasted." By that time, it's often too late.

The problem is, many drinkers refuse to consider drinking as a "disease," which prevents them from seeking help. When people discover they are in the early stages of a "disease," such as cardiovascular disease, cancer, type II diabetes, or other curable ailments, it usually motivates them to act. "Praise God," they think, "that this was caught early. I am in control of doing something about it before it takes my life, and I will get the medical help I need—now!"

By contrast, drinkers don't, and often can't, think this way because, as you learned earlier, drinking affects the brain's rational center. They deny they have a problem, and therefore refuse to change their behavior or get medical help, as they would with any other disease. In the meantime, their brain, liver, and other organs are getting sicker.

MOTIVATING YOUR LOVED ONE TO DIAL DOWN DRINKING–SUGGESTIONS FOR STAYING SOBER

The alcohol effect is different for every drinker, and so is the transformation program required. Here are some general tips to personalize a program that transforms your drinker:

Be realistic. An "alcoholic" has lost control over their drinking, and you are fooling yourself if you believe a person with alcohol disease can stop drinking if they want to. No, they cannot stop. They've lost control. Their desire to drink overpowers their "willpower" not to. Alcohol is an addicting drug. Never dismiss alcoholism as a disease of weak-willed people; try to separate the individual from the addiction/behavior. Have compassion that those who have crossed the line into dependence or alcoholism most likely cannot stop on their own.

Make it a family affair. In a sincere, loving, nonjudgmental way, convey from your heart your concern about the seriousness of the problem. Don't be wimpy. Try to push the button that will motivate the drinker to get outside help, such as:

- "Your family needs you."
- "You shined once before . . . you can do it again."
- "You're such a beautiful person . . . you deserve better than this."
- "Please be open to getting help, now. We'll help you."
- "Your children need and love you so much . . . they want you to get help."
- "We don't want you to wind up paralyzed for the rest of your life . . . or worse, lose your life."

Identify a motivational trigger. It could be something as serious as, *I've got alcohol disease, and now I'm pregnant*. Because of the devastating effects of alcohol on the baby's rapidly developing brain, sometimes pregnancy will motivate a mother to get immediate help. But sometimes the motivational trigger can be as simple as the plea of a young child: "Mommy, please stop drinking . . . it's killing you . . . I need you to watch me graduate from high school . . . walk down the aisle at my wedding . . . and hold your grandchildren in your loving arms . . ."

Recover into something positive. If the person already has an addictive personality, channel their negative, life-threatening addiction into a positive, life-enriching one. This is why we don't call recovering from AD rehabilitation, we call it *transformation*—an inside job. As Dr. Bill's daughter, Erin, recovered from a series of addictions, including alcohol and eating binges, a helpful motivation for her was to turn the problem into a project and experience what the Sears family prizes as the helper's high. She coauthored Dr. Bill's book, *The Dr. Sears T5 Wellness Plan: Transform Your Mind and Body, Five Changes in Five Weeks*, and has an online counseling program, Fit Plus 5 (fitplus5.com), to help those in need transform. Her transformation has been one of the biggest blessings to their family.

A smart and successful family friend of the Searses—let's call him John—slowly crept into alcohol disease. He felt ashamed and estranged

from his two adult children. Although he thought he was hiding his addiction, his children knew, and rightly diagnosed him with alcohol disease.

"John," Dr. Bill counseled, "your kids love and need you. If you get help now, down the road they are more likely to remember how hard Dad fought to stop drinking rather than dwell on Dad being a drinker." John got it.

WHEN DRINKING REGRESSES FROM SMART USE TO ABUSE—HOW TO TELL?

To help you know when you or a loved one needs help, we've identified three progressive stages of alcohol use:

Stage 1: *I drink smart.* You enjoy the taste of an occasional glass of wine paired with a favorite dinner treat, such as a cabernet with a weekly steak. You might enjoy a drink or two at celebrations, but "know when to say when."

Stage 2: *I must have my daily drink.* If you cannot willingly go a few days or a week without a drink, you are in the dangerous stage of *alcohol dependency*, a term we especially use when confronting a friend who would get angry hearing the term *alcoholic*. This stage is an early red flag alerting you to change your drinking habits before your dependency becomes a disease.

Stage 3: *Alcohol disease.* If you *cannot not drink*, are suffering falls and/or blackouts, and/or are doing scary, risky things while under the influence, you have alcohol disease. This means you are likely to be suffering from varying degrees of multi-organ damage, as vividly shown on page 300.

If you think you or someone you love may need help for alcohol disease, we recommend the following groups:

• National Institute on Alcohol Abuse (niaaa.nih.gov)

- Al-Anon (al-anon.org)
- Alateen (al-anon.org/for-members/group-resources/alateen)
- ACOA (Adult Children of Alcoholics) (adultchildren.org)

A CONCLUDING THOUGHT ABOUT DRINKING SMART

Let's imagine that because alcohol is a drug, every bottle of booze came with a package insert or black box warning. Here's how it might read:

> *Do not take on an empty stomach. Sip slowly—with meals. Do not exceed the maximum dosage: two drinks a day for men, one for women. Do not take this drug if you are pregnant, breastfeeding, parenting, or supervising the care of young children. Do not take this drug before driving. Do not take this drug if you suffer from heartburn or digestive issues, are obese, or are receiving psychiatric or psychological care. Do not take with other medications such as antidepressants, benzodiazepines, psychotropics, opioids, or any other mood-altering medication.*

A top message in our brain-health lectures is that one of the best investments you can give your family is to keep your brain healthy. Drinking smart will help that wish come true.

PART III

MORE QUESTIONS YOU MAY HAVE ABOUT BRAIN HEALTH

PEDIATRIC AIR POLLUTION AND INHALATIONAL ALZHEIMER'S

As a parent and part of a big family, I'm so concerned about how air pollution and other toxins that we breathe can affect the growing brains of my children and the aging brains of my parents. Help!

The smog that bothers your eyes and triggers coughing in your lungs may also bother your brain. While air quality can affect the brain health of people at all ages, it is most serious in the very young, whose growing brains are more vulnerable to pollution, and seniors, whose brains are losing their ability to protect themselves against pollution. Brain illness caused by air pollution is such a concern among neuroscientists that they have a scary name for it: *inhalational Alzheimer's*. In children, they often use the term *pediatric air pollution*.

Air pollution neuroscientists, called *neurotoxicologists*, have done fascinating studies showing how fast what we inhale through our nose can get into the brain. To test this, they injected radioactive pollutants into the nostrils of experimental animals, and found that the pollutants enter a highway between the nose and the center of the brain that regulates smell—called the *olfactory pathway*—and from there seep into other highways and harm other areas of the brain.

In 2014, the National Institutes of Environmental Health Sciences sponsored a brain-health workshop where neurotoxicologists were asked to give updates on how air pollution might affect brain health. Their conclusion was that all people, especially children, exposed to high levels of air pollution (defined by the number of brain-tissue-damaging particles in the air) demonstrated:

- More birth defects and premature births
- Slower neurological and psychological development in the first four years
- Decreased intellectual development
- Lower test scores
- More ADHD
- More school absences from respiratory illnesses
- Increased incidence of stroke
- More neurodegenerative diseases, especially Parkinson's

The issue underlying all of these effects is neuroinflammation, which increases brain-tissue damage and slows brain development and thinking. Because the brain is the most precious organ of the body, it is flooded with neuroprotectants. The brain's resident immune cells are called *microglia*, and trillions of these microglia constantly patrol the brain, trying to clean up any pollution. Yet, after years of exposure to pollutants, this built-in protective system becomes overwhelmed. Like outnumbered soldiers, the microglia can't keep up—though they continue to try, which can cause a hyperactive inflammatory response. The pollutants take over, causing neurotoxicity.

The effects of air pollution on the brain get worse. As neuroscientists continue to study air pollution's impact, they are finding that many of the brain cells that fertilize the growing brain garden are being damaged by air pollution. Two such cells of concern are called *astrocytes* and *oligodendrocytes* (the cells you saw in action in "Your Brain Garden Parts" on page 11). Simply put, pollution damages the seeds in the growing brain garden.

Pollution also causes brain leaks. Because air pollution can damage the astroglia, the cells of the BBB that repair this brain protector, continued exposure to air pollution can break the BBB down, which further exposes the brain to pollution.

Some simple steps you can take to have greener air:

- Try not to live or go to school near to or downwind from a freeway.
- Switch lanes to avoid breathing diesel exhaust from the truck in front of you. Little breaths can add up to a lot of brain damage. There's even a name for this road-based exhaust: TRAP (traffic-related air pollution). While driving, if you can see it or smell it, try not to breathe it.
- Certainly, keep your child away from smokers, since secondhand smoke can damage lungs and brains. We take roughly ten thousand breaths per day, which is a lot of opportunity for pollution to enter the lungs.
- Lobby for better air-pollution laws. While many good laws have been passed and air is certainly getting cleaner in many cities, there still are a few loopholes: big trucks spewing out fumes of diesel fuel still go unregulated, for example.

In 1992, the University of Southern California (where Dr. Bill once served as assistant professor of pediatrics) launched the Children's Health Study, in which researchers collected data from more than six thousand children throughout southern California and compared their health with the levels of air pollution over the next eight years. Not surprisingly, they found that children who lived closest to the most polluted air got the most respiratory illnesses, such as asthma. And we now know that if polluted air bothers the lungs, it is likely to bother the brain, too.

WHY CHILDREN ARE SO VULNERABLE TO DIRTY AIR

Not surprisingly, neurotoxicologists discovered that the younger the child and the more polluted the air they breathe, the greater the risk of delayed brain development. Growing brains are particularly vulnerable for these reasons:

- Children have a larger lung surface area in relation to their total body weight, and breathe around twice as much air per pound of body weight compared with adults.
- The genetic machinery of growing cells is more sensitive to toxic chemicals.
- Children spend more time outdoors, and therefore breathe more polluted air, compared to adults.

Yes, "go outside and play" is one of the best medicines for growing brains, but they should play in clean air.

A GREEN ENVIRONMENT

Our family is trying its best to live green. Any suggestions?

It's no medical mystery why we have an epidemic of autism in kids, autoimmune diseases in young adults, and a huge increase in Alzheimer's disease in seniors, beginning at a younger age. We are polluting our brains with toxins, every hour of every day. A classic study by the

Environmental Working Group revealed more than two hundred toxic industrial chemicals and pollutants in umbilical cord blood from babies born in US hospitals. And as a sad reflection of how dirty air dirties growing brains, a study revealed that some children living in heavily polluted Mexico City already had beta amyloid plaques (an indicator of pre-Alzheimer's) in their brains.

While we all have to live in the world we have, here are ways you can treat your brain to a greener version of it.

To fix your brain, first fix your food. The smart eating plan you learned in chapters two and three is not only good for the brain, it's good for your body's garbage-disposal systems: the liver, intestines, kidneys, and so on. Most people's garbage-disposal systems are overwhelmed by their eating and breathing more toxins than they can handle. If you avoid the garbage you *can* control—in the food you eat—your body will be better able to handle the garbage you can't control. And the more smart, immune-strengthening foods you preload your body with, the more pollutant-fighting soldiers you'll have.

Sleep in a green bedroom. First, check what you sleep on. If your sheets, pillow cases, and blankets are not made of organic cotton, suspect they include *organohalogens*, or chemical fire retardants. Are there fire retardants in your mattress? Is the artificially made wood on your bed filled with *formaldehyde*?

If your bed isn't bothering your brain, what about the air? Are you breathing in fumes from chemically made furniture and carpet (the chemical that allegedly protects your carpet from stains may not protect your brain from "stains"), or exhaust from outside air or a power plant just a few miles away? You spend more time in your bedroom than anywhere else. Make sure the air is clean by sleeping in a bed and room free of the pollutants just mentioned.

Check your bathroom for toxins. Some personal care products—sunscreen, hairspray, cosmetics, soaps, antiperspirant, nail polish—may not be good for brain health, especially if they may contain bisphenol A (BPA), a possible neurotoxin. To find out what personal care and household cleaning products are safe and which are toxic, see the lists at EWG.org.

Also, check your water for chlorine, which when inhaled can damage the protective lining of your airway. Purchasing a filter for your shower head protects you from chlorine exposure.

Beware of mold. Mold spores release something called mycotoxins, which can overwhelm the immune system. If you or your children have frequent nasal and sinus congestion, get your home checked by a certified mold specialist. To Dr. Bill's surprise, when he had his house checked, their attic was a moldy mess. Cleaning it up made a big difference in his family's health.

~~~~~~~~~~~~~~~~~~~~~~

When you change your own behavior, you can also help make a difference on a societal level. The EPA justifies "acceptable levels" of neurotoxins in products by saying they contain "only trace amounts." While that may be true, neurotoxins also have a *cumulative* effect. These "safe" microexposures gradually overwhelm the body and the brain's garbage-disposal system. After so many years of exposure, it's no wonder the fifty-year-old brain says, "Enough already. I've had it. I'm on my way out!" If you, the consumer, refuse to buy anything that contains neurotoxins, the dirty money will slow down, and less pollutants will be dumped into our food and air—treating us to cleaner blood vessels and healthier brains.

ENDOCRINE DISRUPTORS

I've heard a lot about endocrine disruptors affecting the brain. What does that mean?

Do you have *endocrine-receptor dysfunction*? The term "receptor" is one that has crept into many smart books and articles on brain health over the past few years. Put simply, a receptor is a site on a cell that "receives" and takes in a particular hormone. Endocrine-receptor dysfunction means the appropriate hormone is not entering the cell, impairing cell function.

Picture cell membranes as having electrochemically controlled doors. When a helpful neurochemical, such as a molecule of thyroid hormone, arrives at a door, it provides a password. If the hormone's password matches the door's, the door opens and the hormone gets sucked into the cell, where it can do its work. Each hormone has its own personal receptor sites on a cell.

However, artificial chemicals—appropriately dubbed *endocrine disruptors*—can block these doors from opening or confuse the passwords, disrupting our normal biochemistry. The natural neurochemicals can't get in. Simply put, these chemical toxins disrupt your natural endocrine hormones' ability to do their healthful job. Common endocrine disruptors are pesticides, plasticizers like bisphenol S, and flame retardants.

More chemical disruptors. Toxic chemicals like pesticides that creep into our gut-brain lining can also act as *junction disruptors* because they can cause injury to the tight junctions between the gut cells, leading to leaky gut.

Endocrine disruptors trigger autoimmune diseases by damaging healthy tissue, such as the thyroid. The hyped-up immune system then attacks the damaged tissue, mistakenly thinking it is a foreign molecule. And conditions like autoimmune thyroiditis, also called Hashimoto's disease, can harm your brain health; as you learned on page 216, the brain and the thyroid are partners in mental wellness.

HEAVY METAL TESTING

I've heard a lot about heavy metal testing for possible brain problems. Tell me more about it.

Remember, one of the themes of this book is to think outside the box—that is, think outside your brain. Many brain-health issues can be triggered by exposure to environmental toxins such as heavy metals. In our medical practices we have seen patients whose brain health markedly improved following chelation therapy, which pulls heavy metals out of fat-cell storage and bones, into the blood, and then out in the urine.

A chelation specialist will first take your history of most likely exposures: "Do you work in an improperly ventilated hair salon?" "Do you live near industrial pollution?" and so on. Next, a blood, urine, and/or hair analysis may be performed. Then, appropriate therapy may be recommended to inject "chelators," such as EDTA, intravenously to bind the toxins so you can excrete them.

Here's a story from one of Dr. Bill's patients:

In the summer of 2015, my oldest son came to me in sheer desperation and said, "Mom, I don't think I can go on anymore feeling like this." He was struggling with very serious anxiety and depression. As a mom, all you can think is, "How can I help my child?"

Well, just a few weeks later—I totally believe it was a God-ordained appointment—I met an amazing woman who is a healthcare professional in Illinois and I learned she helped people get rid of heavy metals and other toxicities that are so prevalent in our environment.

The first thing we did, per her request, was have some new tests run, because the regular tests that the doctors ran before didn't show anything, and we knew something was wrong. She ran about fourteen different blood tests, including for C-reactive protein, homocysteine, thyroid, T3, T4, and more.

The first thing that came back, which was a huge red flag, was that the C-reactive protein levels, which indicate inflammation, were off the charts. The next thing that came back was a hair analysis. This was the most telling. The hair analysis showed that he was off the charts in uranium. She asked my son, "Have you ever been to Russia or Japan or have you served in the military?" He said, "No." I said, "Where are you going with this? Do you mean like nuclear power?" She said, "Yes, I only see this high level of uranium in people who have been exposed to either the Iraq War or a nuclear power plant."

I started thinking. "Oh my gosh," I said. "We raised our family in Southern California right near the San Onofre Nuclear Power Plant." How else could he have such high levels of uranium? And my son had been a surfer since he was about twelve years old and was in the water every single day right in front of the nuclear power plant.

Then more tests were done, through urine and saliva, that showed that his neurotransmitters and his hormones were off. From further research, as well as a consultation with our healthcare practitioner, I learned neurotransmitters can be affected by heavy metals.

This savvy mom just didn't settle for the "take this" pills approach to her son's depression. Instead she searched for the root cause and treated that. With expert treatment, the whole family's uranium levels were drastically reduced, and her son's depression was completely healed.

BLOOD TESTS FOR BRAIN HEALTH

Are there certain tests my doctor may do for my brain health?

Yes. Doctors use blood or urine tests for what they call "biomarkers" to find out whether something is out of whack in some system of your body, including the brain. What's important to know is that some biomarkers go up in your blood long before you feel the effects in your body. This is especially true for brain disorders.

An analogy we have used for years is that having regular blood tests done is like keeping an eye on your car's dashboard warning lights. When the "check engine" light goes on, it alerts you not necessarily to stop and call a tow truck, but rather to make an appointment at your earliest convenience with your friendly neighborhood "car doctor."

An astute dashboard-watcher, analogous to a preventive-medicine advocate, watches the gauges every day. They notice that the temperature gauge mostly stays on normal. One day, while driving to work, they see the gauge has inched up slightly from its normal range. This smart person then pulls into the nearest car doctor's office to get the meaning checked, treat the root cause, and drive off with a healthier car.

Mr. or Mrs. Put-It-Off, on the other hand, waits until the gauge goes all the way over to the highest temperature and they notice smoke coming out of the engine. By this time, engine damage, sometimes irreversible, has already been done. The car now has a "disease" that could have been prevented if it was treated in the "dysfunction" stage.

Things your healthcare provider may run tests on to keep an eye on your brain health include:

- Biomarkers for inflammation and autoimmune diseases: antinuclear antibodies (ANA), ESR, and/or C-reactive protein
- Blood omega profile, especially red-blood-cell omega-3 index and omega-3/omega-6 ratio (see explanation, page 45)
- Homocysteine, a marker for increased oxidative stress, which is highly correlated with Alzheimer's disease
- Blood nutrient levels of B_{12}, folic acid, vitamin D, iron, and zinc

Your doctor should also perform:

- A complete metabolic profile
- A thyroid profile (hyperthyroidism or hypothyroidism can resemble early Alzheimer's and trigger depression; be sure thyroid antibodies are included in the test)

KETOGENIC DIETS

I've heard that a ketogenic diet can help people with some brain-health problems such as Alzheimer's. How does it work?

Dr. Bill's interest in ketogenic diets (KDs) began in the early seventies when, as an associate ward chief of a neonatal intensive care unit, he realized that most babies can thrive on "ketones," an alternative fuel produced by that adorable excess baby fat. Breast milk is also particularly high in smart fats called MCTs (see their description, page 57), which are metabolized into ketones. That's smart protective design! He was even more impressed when he learned about ketones' cancer-healing perk. Cancer cells thrive on glucose, but not on ketones. (See Dr. Bill's cancer survival story, page 169.)

Pediatricians have been using ketogenic diets for more than forty years to treat children with epilepsy, and it has long been known that in carefully selected people with certain neurological problems, going ketogenic definitely does help.

Ketogenic diets and their milder cousin, *intermittent fasting*, have become much more popular in the last few years (although even in biblical days "fasting" was high on the list of healthy habits). But you may be surprised to hear that you already "intermittently fast" every day during the night. If you don't eat after an early dinner and then "break the fast" with a breakfast around 7 AM, that's twelve hours of fasting. Presently, there is no official medical definition for intermittent fasting, because there are so many ways to do it: twelve hours at a time, a full day of water only, and so on.

Here's what happens when you fast and why it can be healthy for many people. When you fast, your body runs out of sugar. This can occur as soon as twelve hours after eating because many people store only twelve hours' worth of sugar fuel (glycogen) in their liver. When you run out of sugar, your internal fuel gauge sends a biochemical text message to your fat cells to empty excess calories for fuel to keep your brain and body going. Your fat cells then send those fats to the liver, which converts these excess fats to *ketones*, a *neuroprotectant*. A "keto diet" is high in healthy fats, medium in proteins, and low to very low in carbs. Consider ketones as a super-fuel for the brain, like a super-grade gasoline. As fuel, ketones are also exceptionally clean-burning: unlike glucose, ketones don't produce as much exhaust or *oxidants*, that wear-and-tear word you learned in chapter one. More power, but less exhaust—brains like!

WHY KD WORKS TO PREVENT AD

As we age, the brain's ability to burn glucose lessens to varying degrees. So, to prevent dementia, the smart brain welcomes a backup fuel it likes: ketones. In fact, one of the emerging theories about AD is that it is a dysfunction of glucose usage, and unless the brain tissue can turn to an alternative fuel, it just wears out. The aging brain continues its ability to use ketones, even as glucose metabolism lessens.

Alzheimer's is correctly called "type III diabetes" because, as part of the disease, certain areas in the brains of Alzheimer's patients develop insulin resistance. That means these areas don't efficiently use their preferred fuel, glucose. When brain tissue doesn't take up enough fuel, it doesn't perform correctly. A ketogenic diet provides a fuel replacement.

There are two ways to access ketones as fuel. First, you can make your own by going on a strict ketogenic diet. Our bodies have a natural built-in fuel bank of ketone reserves that our hypermetabolizing brains can quickly dip into when needed. That bank is body fat. Picture the brain texting the extra body fat, especially in the belly, buttocks, and thighs, and saying, "We're running out of glucose up here. Quick, burn some of that extra fat and send us up some ketones!"

Your keto specialist can advise you of your options and help you find the right ketogenic diet for you. For example, if you are a younger person or you are not showing any signs of early dementia or MCI, then simply eating a daily tablespoon of coconut oil as preventive medicine may be enough. Yet for people who are suffering from MCI or early dementia, or who are already diagnosed with Alzheimer's, the keto specialist may prescribe both a ketogenic diet and daily coconut oil or MCT oil supplements. (Notice that our supersmart smoothie on page 73 contains coconut oil.)

HOW TO GO KETO SMARTLY

Suppose you or a family member shows early signs of dementia or AD. Because KD requires a major dietary change and one formula does not fit all, follow these steps as you make the shift:

1. **Consult a specialist in ketogenic diets.** You can find them online at charliefoundation.org and ketogenic-diet-resource.org. Your keto specialist will make sure there are no reasons why you shouldn't go keto, such as if you are on certain medications or have liver disease. You will also be given a list of things you should carefully monitor, such as your blood sugar, since sometimes, as the brain shifts to preferring ketones as its primary fuel, your blood sugar may drop below healthful levels.

2. **Personalize your KD.** You may need a modified KD: a not-so-low-carb diet, intermittent fasting, or a very-low-carb diet. Obviously, for the ketones to take over as the primary fuel, you are going to have to drastically reduce the carbs in your diet, but go slow.

   A modified KD we find that fits many people is:

   - High in healthy fats
   - Medium in proteins
   - Very low in fiberless carbs, "added sugar," and processed carbs
   - Rich in fiber-packed chewy vegetables

   The healthy brain diet that we advise in chapters two and three fits these four smart changes. One of the reasons a nutritionally messed-up brain will like this modified keto diet is that our healthy brain diet *replaces unhealthy carbs with healthy fats*.

3. **Don't forget to feed your gut brain, too.** The discovery of the mind–microbe connection you learned about in chapter three prompted recent concerns that a strict, unsupervised KD may not be healthy for the gut brain. The main reason is that gut bug diversity can be low in some people who adhere to a very low-carb diet. One solution: Eat more vegetables. Yes, vegetables are carbs, but while they're high carb on paper, they usually aren't so in your bloodstream. You burn most of the carb calories in most vegetables just by chewing and digesting them. And they are also rich sources of the microbiome's smartest food—fiber. Enjoy low-carb/high-fiber vegetables such as artichokes, greens, celery, kale, spinach, and chard. (See list of fiber-rich favorite foods, page 89.)

4. **Stay rich in antioxidants.** Removing all fruits, as some KD diets do, can be risky because they are full of the brain-saving antioxidants you learned about on page 13.

5. **Go slow with MCT oil.** Some people experience diarrhea when eating too much MCT oil too fast, so some specialists

recommend starting with a half teaspoon of MCT oil (or one teaspoon of coconut oil) once or twice a day, and gradually increasing to one to two tablespoons.

6. **Drink more water.** Since some ketogenic diets can have a diuretic effect (a.k.a. you pee more often), drinking more water can keep you from getting dehydrated.

Following these sensible and science-based steps will help you personalize the KD that's best for your brain.

VEGAN BRAIN

Is going vegan healthy for my brain?

Yes, providing you do it smartly. Some of our healthiest and smartest doctor friends are vegan (absolutely no animal foods), but they do it right by supplementing with nutrients they need but that a strictly vegan diet doesn't sufficiently provide. However, for many eaters, going totally vegan is risky.

While writing this book, Dr. Bill was talking to eleven-year-old Gracie, who proudly announced, "I'm vegan!" Before he and Gracie's doctor-dad could launch into our sermons, she added, "Because I can't eat anything with a soul!" They quickly took off their doctor hats, put on their grandfather and father hats, and hugged her for being such a kind and caring person.

It's interesting, and sometimes concerning, that veganism is on the rise. Dr. Bill recently attended a high-profile medical meeting of top scientists. What was for breakfast? Tofu scrambled eggs. Yuck! Although selective scientific studies may support a vegan diet, many studies, and nearly all scientists, agree the Mediterranean and Asian diets are some of the brain-healthiest in the world. Why? These diets are primarily *pesco-vegetarian*—vegetarian plus seafood.

Three reasons why going vegan could be smart, theoretically:

1. Vegetables are one of the smartest sources of *antioxidants*. As you learned on page 13, the brain thrives on antioxidants.
2. Vegans tend to be *leaner*. As you learned on page 103, leanness partners with smartness.
3. Vegans have a much lower risk of cardiovascular disease, as my friend, mentor, and one of the most respected preventive-medicine specialists in the world, Dr. Dean Ornish, proved in his scientific studies on the blood-vessel health of people who want to prevent or reverse cardiovascular disease (see his highly recommended book *Dr. Dean Ornish's Program for Reversing Heart Disease*). And Brain Health 101, as you learned on page 10, is: the healthier your blood vessels, the healthier your brain.

However, going vegan may not be smart for everyone. Consider this big "fat" problem: vegan diets are usually "low-fat," and even worse, low in smart fats. As we have seen, your brain is a "fathead" and requires smart fats to thrive. Additionally, many of the smartest nutrients the brain needs are low in vegan diets, such as DHA, iodine, iron, selenium, copper, zinc, vitamin D, some B vitamins, choline, and other nutrients.

If you choose to go totally vegan, do it smartly:

- Visit a functional medicine specialist periodically and get blood tests for nutrient levels that may be missing in your vegan diet, including your blood levels of the smartest fats (DHA and EPA) and your omega-3/omega-6 balance. (See Vital Omega-3/6 HUFA Test, page 45.)
- Take smart nutritional supplements to bridge nutritional gaps. Readers, remember: The smartest brain fat is DHA, and DHA is only found in *seafood*, not land food. (See more, page 32.) Those who want to go strictly vegan but still eat smart *seafood* should take a seafood supplement that comes from algae, the sea plant that fish eat to get their smart fats (see AskDrSears.com/supplements).

Our advice for most people: go *pesco-vegetarian*. Eat primarily vegan, but eat safe seafood at least twice weekly, an egg a day, wild game occasionally, and a bit of organic pasture-raised whole milk and dairy. While there is likely always going to be controversy over which is the smartest brain-health diet for most people, the current leader is the Mediterranean diet, which is rich in the smart foods we recommend in chapter two.

"GLUTEN BRAIN"

So many of my friends are going gluten free, but I suffer from gluten confusion. How much is hype, and is gluten really bad for your brain?

Like most grains, real wheat is good for the brain. It's what the food industry has done to wheat—or rather, to gluten, the "glue" in wheat that keeps the dough together—that makes it bad for the brain. Hybridization has created modern wheat products with very high levels of gluten proteins. When these high amounts of gluten get into the gut, they are broken down into smaller protein chains, one of which is gliadin, the part of gluten that some guts simply don't like and react to.

The lining of your gut brain is very protective and selective. Suppose some wheat approaches the lining and wants to get through. The smart gut brain says, "Oh no, you're not real wheat. You've been molecularly messed up. We don't recognize you as a real food, so we're going to fight you to protect our partner, the head brain." The immune system then mobilizes its army to fight the unreal food. The molecular mess from this battle inflames the tight junctions of the gut lining, widens them, and contributes to a leaky gut. And as you learned in chapter three, a leaky gut can lead to a leaky brain.

Some people experience what we call *gluten brain*, mental fatigue and brain fog after eating wheat. Also, in some cases, the immune system can get messed up, *over*fight, and trigger autoimmune illnesses.

In deciding whether or not to go gluten free, we recommend: *when in doubt, leave it out!* Also, keep in mind the following:

- Rely more on common sense and your symptoms than on blood tests to know if you are gluten intolerant. Don't wait until inflammatory markers in your bloodstream show you have a leaky gut or until you get pains in the gut. Again, use common sense. If it's not real wheat, don't eat it. If you already have brain-health issues or want to prevent them, we suggest you go gluten free. A wise resource to help you is *Grain Brain* by Dr. David Perlmutter.
- Just because the label says "gluten free" does not automatically mean it's a healthy food. Many gluten-free products are too low in the smart nutrients protein and fiber, too high in added sugar, and not nutrient dense, meaning they pack less nutrition per calorie than good old-fashioned made-with-real-wheat foods.
- Foods made with "whole grains" are healthier than those made with white flour because whole grains still contain one of the healthiest parts of the wheat, the wheat germ. In white flour, this nutrient is removed because keeping the wheat germ makes the flour turn rancid and get moldy quicker, and therefore has a shorter shelf life. If you are avoiding gluten, consider healthy whole-grain alternatives, such as quinoa, amaranth, and wild rice.

IS DAIRY SMART?

I've read a lot of articles bad-mouthing dairy and I'm confused. Help! Is dairy smart or not?

Most infants and young children, especially in the preschool years, have enough lactase to digest lactose, but after that, more than half of humans gradually lose some or much of their lactase and become dairy intolerant. Because of this, could we assume that dairy is healthy in the preschool years and then gradually should be phased out of the diet between five and ten years of age? Dairy food for thought.

However, we believe dairy can be smart, especially if you get it, and drink it, right from the source—the goat or the cow. Picture being a farmer at milking time. You go out to relieve your cows and goats of their full udders. The milk you get is smart because of the fat it contains and even the bacteria that your gut brain likes. But then the milk gets pasteurized, which can mess up the healthy *Lactobacillus* bacteria that your microbiome likes, and homogenized, which lessens the digestibility of the fat.

If you like eating and drinking dairy, do it smart by making sure your dairy is:

- Organic
- From 100 percent grass-fed animals
- "Whole" (versus nonfat)
- Plain (no added sugar or artificial sweeteners)

Also, consider dairy that has been fermented, like yogurt, kefir, and cheese. Fermenting lowers lactose, making dairy more digestible. And goat milk may be smarter for your brain, since it is higher in the smart fat MCT (see page 57) than cow milk.

Finally, to give your gut a fighting chance to digest most of the dairy, sip it slowly with meals, as in our supersmart smoothie (see page 73).

GROW YOUR GARDEN, GROW YOUR BRAIN

I've heard that gardening is good for the brain. Is this true? Why?

It's true! When you garden, you're outdoors in natural sunlight, moving and creating, and later you get to eat the smart foods that your labor lovingly produced. These gardening perks explain why neuroscientists have shown that people who enjoy gardening are less likely to suffer Alzheimer's.

Studies also showed that gardening changes your brain, often for life. Schoolchildren who participated in gardening were more likely to become adults who enjoyed gardening. And, as Harvard neurologist Eva Selhub mentions in the book she coauthored, *Your Brain On Nature*, children who grew up gardening were more likely to value trees as a visual delight to improve their mental health. (For a fun and family-friendly way to create a home garden, see AskDrSears.com/tower-garden.)

DESIGNING A GREENER OFFICE

I love my job, yet my brain is not as happy as it could be in my office. What can I do to make my office more enjoyable to work in?

The more smartly you design your workplace, the smarter your brain works. Employers, got that? And one of the quickest and easiest ways to make your brain less stressed and happier in the workplace is to bring the outside in.

Plants. Nature neuroscientists have discovered that not only does green space promote peace, but the greener your office, the greater your ability to relax, think, and create. Viewing plants in or from your office increases productivity, decreases depression and anxiety, and reduces sick leave.

Windows "wow!" the brain. Let the daylight and sunlight into your brain. With artificial light, you're more likely to get artificial thinking. Your brain lights up when you let the sunshine in. (See page 271 on studies showing that students remember less when they read information in artificial light.)

DR. BILL'S OFFICE DESIGN STORY

In 2006, I began studying the therapeutic value of sunlight, especially in the workplace. At that time, we had the pleasure of designing our new office building for Sears Family Pediatrics. Our design feature #1: let the healing sunshine in. Windows surround our waiting room, a skylight umbrellas the central office area, and in each exam room the doctor's desk faces the window. Smart!

What's true for office buildings and hospitals is also true for designing a smart classroom. A study of more than two thousand students showed that those who enjoyed the largest classroom windows learned better, especially in reading and math. And Japanese neuroscientists showed that greening a high school classroom with plants decreased the number of student sick days. Japanese medical specialists have also showed that patients whose beds faced the window healed better. Yes, sunlight is good medicine for the body and brain.

iBRAIN

I'm concerned about the brain effects of all this technology. How can I use it to help my brain instead of harming it?

In many ways, technology has made our lives easier, happier, and less stressful. It's reassuring to have GPS remove travel worries by telling you where you need to go. Text messages and photo-sharing on smartphones help keep families connected. Yet, like with prescription medicine, while the right dose may help, an overdose may hinder.

Dr. Bill likes to tell this story: One afternoon during a family birthday celebration, he put this sign next to their door, to be viewed and heeded by all who entered their home: "Welcome to our Fun Home! Please deposit cell phones here."

Of course, he got a few looks of *Oh, grandpa . . .* and a few raised eyebrows from the cell phone-attached adults. The adults seemed to have more of a problem with his wish than their children; after a few minutes of imagined boredom (mother of eight, Martha, always told our children: "Boredom is a choice!"), the kids were really getting into the fun and games. They were enjoying camaraderie instead of cell phones and dinner-table talk instead of screen-checking. That day they got out of themselves and enjoyed others.

We and nearly every other healthcare provider are intrigued by what we consider the most concerning experiment on human brains to ever affect mankind: the overuse and abuse of screen time. An important concept we mention throughout this book is *balance*, which applies also to the smart use of screen time. Here's a summary of our concerns, especially what science says about screen-time addiction:

More metabolic syndrome. There is a dose-dependent association between metabolic syndrome (prediabetes and obesity) and screen time in adolescents. The term "dose-dependent" means that the more time you spend in front of a screen, the more excess fat you accumulate throughout your body.

Increasing social anxiety. To get along with the smartphone crowd, young adults feel they need to go along. As a result, they are becoming more comfortable relating to screens and more uncomfortable relating eye to eye and face to face with real people. Social anxiety is another modern malady that has entered the psychological anxiety list. Its increased incidence parallels increased cell phone use. Any connection? We believe yes.

More "text neck." In our medical practices we see many humped-over teens and young adults looking more like older folks. This poor posture is especially worrisome in younger teens, whose backs are growing rapidly, but it can lead to major back problems at all ages. To balance "text neck," we recommend trying to spend almost as much time looking up as you do looking down, such as swimming the breast stroke, or saying "hi to the sky" (as you extend your neck and enjoy eye-pleasing sky colors) often throughout the day. Also, hold your cell phone at eye level as much as possible.

Delayed language, social, and motor development in young children. A lead article reported in the March 2019 issue of *JAMA Pediatrics* studied more than three thousand families with children two, three, and five years of age who engaged in an average of two to three hours of screen time per day. The main takeaway of this study was that the amount of delay in language, social, and motor development was directly proportional to the amount of screen time. In addition to the researchers' conclusion, the concern we'd add is about what these children are missing during this screen time, one of the most important contributors to childhood development: namely, *go outside and play*.

Harm to visual development in young people. If screen time addiction continues, neuroscientists predict that 75 percent of adolescents will wind up needing to wear glasses. Spending too much time too close to artificial screens causes the eyeball to grow too long. As a result, images fall in front of the retina rather than directly on it, making the objects blurry and appear farther away than they really are. This is called myopia, and it typically starts in childhood or adolescence and progresses into the mid-twenties when the eyes—and the brain—are fully grown. Once upon a time, eye researchers attributed myopia to genetics, yet new insights have shifted the cause to increasing time spent staring at screens and exposing the eyes to artificial light all day. Exciting new research is also revealing how natural sunlight is the preventive medicine for these eye problems. Imagine taking your child in to an eye doctor for prescription glasses and the doctor writes down on a prescription pad: *Spend more time outdoors*.

Disruptive and depressing. We are already a hyped-up generation. Do we need more artificial adrenaline rushes? A screen-induced adrenaline rush may feel good in the moment, but hyper hormones are like gravity—what goes up must come down. The only way to keep them from crashing is constant screen-time use (a.k.a. dependency, or addiction). A smart person in Dr. Bill's medical practice told his friends, "Only text me good news. If it's not good news or you need help, call me so we can talk it out." The more we balance our screen time, the better we balance our brain.

Eye-to-screen instead of eye-to-eye contact. The screen does not (usually) smile back or talk back. Our concern, in a nutshell, is that because screen time overusers and abusers get less good old face-to-face contact, they are more likely to fall into a state of chronic stress and anxiety. Also, scientists who study people addicted to brain-bothering screen content find that they score lower on tests of emotional intelligence, and especially show less empathy. Some habitual screen-watchers are so used to interacting with screens that they have difficulty interpreting facial expressions and cues. Next time you're in a crowd, watch how those people whose eyes and minds are magnetized to the screen of their cell phones are seemingly oblivious to the social interactions going on around them. Next time you're in a restaurant, notice how many people are sitting at a table and are engaged more in their i-something than in the faces of those around them.

Is screen time really an addiction? Many times, yes! Researchers studied this by asking some college students not to use their cell phones for a day or so. Many suffered from withdrawal and showed symptoms of anxiety and depression. The findings were very similar to what you see in people being weaned off a drug addiction.

Do screens harm learning? Some education specialists believe yes (which is alarming, given that, in some schools, iPads are replacing books and the printed word). In studying how people learn, researchers have noticed that memory attention is greater when studying the printed word in a textbook than when read on a screen. This makes sense, since the artificial colors and lights on a screen are more likely to tire the eyes, and tired eyes lead to a tired brain.

Has cell phone safety been proven? Not really. Studies alleging "no link between the electromagnetic field of cell phones and brain cancer" are short-term studies, just like the early tobacco and lung cancer studies that showed no correlation. It may take ten to fifteen years or more for brain cancer to develop. So we don't yet know about cell phone safety, especially for young children and teens, whose brains are growing and changing so fast. And what we have learned is troubling: The massive Interphone Study in 2008 revealed that the risk of a brain tumor increases on the side of the head against which users more commonly hold the phone.

For updates on the smart and safe use of screen time, see AskDrSears .com/screen-time.

CELL PHONE USE SAFETY TIPS

Until more cell phone safety research is done, we suggest these preventive measures:

- Do not let children under twelve use cell phones. (Even Steve Jobs advised not before age fourteen in his own family.)
- Remain at least four inches from your phone while talking by using the speakerphone function or a Bluetooth headset.
- If you must put your phone close to your head, as soon as you make the call, wait for the other person to pick up before placing your phone close to your head; the electromagnetic field is greatest when you begin the call. Also, change ears frequently.
- Keep away from people using their cell phones on trains, buses, or in other tight quarters.
- Don't keep your phone close to you at night.
- Enable airplane mode when your phone is not in use, which stops electromagnetic emissions.
- Avoid using your phone when the signal is weak or during rapid travel, such as by train or car, because during the phone's constant search to connect with a transmission antenna, its power output is at maximum.

THE HEALING POWER OF TOUCH AND HUGS

I love to hug and be hugged. What is it about hugging that makes you feel so good?

Dr. Bill recently chaired a conference in Chicago about the neuroscience of hugging, where neurologists, therapists, and psychologists looked at the effects of hugging throughout the body. One of the main points he made during his lecture was that hugging for twenty seconds puts your body and brain in balance. It calms anxiety, lifts depression, balances your immune system, and lowers high blood pressure. The eye-to-eye contact, touch, and often the bonus sweet words all add up to the feel-good effect of a happy hug. This is why a person who is having a down day often sees a friend and says, "I need a hug!"

Scientists have found that there are changes in brain pattern activity during touch. Touch reduces the stress hormone cortisol and releases oxytocin, known as the "feel good" hormone, helping to inspire positive thinking and an optimistic outlook. It helps humans connect and promotes feel-good sensations that foster a sense of well-being. Physical touch also increases the levels of dopamine and serotonin, two neurotransmitters that help regulate your mood and help your body relieve stress and anxiety. Additionally, physical touch is known to improve your immune system.

Hugging seems to create its well-being effect in the thalamus, the emotional hub of the brain, which then sends "peace be with you" text messages throughout the brain and the body, and is sometimes even partnered with goosebumps. (See "Endnotes" for why you get goosebumps.) Letting a hug last for twenty seconds, science shows, is the time it takes for a hug's feel-good effects to click in.

To better a brain, hug someone today!

COACH HAYDEN TALKS ABOUT HUGS

I was born with a high need for physical touch, along with the temperament that allowed this need to be known. Fortunately, I was born into a high-touch home, and my parents quickly learned that touch was the "medicine" that I needed for my sense of well-being. When left alone for too long without physical connection, I would become agitated, fussy, and often highly distressed.

While it is common to comfort babies and small children physically, the need for physical touch is often ignored in older children and adults. Touch is the most primitive of all senses, the first one to develop, and our need for it does not go away as we age.

Signs you may be suffering from touch hunger include:

- Loneliness
- High stress levels
- Agitation
- Aggressive behavior
- Body image issues
- Depression
- Anxiety
- Feelings of unworthiness or rejection

Ways to incorporate more physical touch into your life include:

- Increase the number of hugs you give and receive
- Intentionally increase the duration of hugs that already occur
- Walk hand in hand with a family member, close friend, or romantic partner
- Take a ballroom dance class
- Get regular professional massages, or trade massages with a close friend

Try to give and get a few hugs every day. The longer your hugs last, the better both you and the person you're hugging will feel. A prolonged, deep hug shows you really care.

TRAUMATIC BRAIN INJURIES

Our sixteen-year-old was hit in the head by a soccer ball, suffered memory loss, and had to be hospitalized for TBI. What can we do?

TBI, also known as a concussion, means the shaken brain is jolted into a state of neuroinflammation or brain tissue stress. The bruised brain may then quickly recover from this shake-up, or it may continue to show subtle or serious effects.

Dr. Bill once suffered a severe TBI from being hit in the head during a medical mission trip to help in a hospital for brain injuries in a developing country. His doctors were amazed at how quickly and completely he healed—it took a month for him to get over the brain fog—and asked him, "What's your secret?" He told them about the importance of preloading your body and brain, so when things like traumatic brain injuries happen, your immune system is ready to quell the neuroinflammation.

Healing 101 is about turning a problem into an opportunity. The first part of this book could be your textbook for improving overall brain health. Read our healthy brain plan and do it with your teen. Next, we recommend picking up a copy of *When Brains Collide*, by Michael Lewis, MD. Dr. Lewis is a friend of Dr. Bill's, as well as a top TBI specialist for the US military, and has had amazing results helping people heal from TBI. Visit Dr. Michael Lewis's website at braincare.center.

~~~~~~

YOUR DAILY-DO LIST TO ENJOY A CALMER, HAPPIER, SMARTER YOU

Remember, the main feature of our healthy brain plan is to teach you skills. A wise healthcare provider holds you accountable: "Did you take your pills as I prescribed?" It's even more important for your therapist or doctor to help you do—and keep doing—as many of the recommended self-help skills as you can. With that skill-enforcement in mind, we've provided a Daily-Do list of skills below, which you can also download at AskDrSears.com/brain-health-daily-do-list.

Use this checklist daily to track your progress toward building a healthier brain. Be honest! You won't get graded on right or wrong answers. Your goal: Each week, the "No—Why?" column should have fewer marks.

If you are initially seeking therapy advice for *skills*, it can help to go through this checklist with your skills coach/therapist. They can then see at a glance what skills you most need help in doing and what personal hurdles you have to overcome.

MY COMMITMENT	Yes	No—Why?
I believe I can balance my brain.		
I believe I can think-change my brain.		
I'm committed to taking charge of my brain health.		
I believe I am the orchestra leader of my own cerebral symphony.		
EAT SMART		
I eat two fistfuls of wild salmon each week.		
I take 1,000 milligrams of a DHA/EPA supplement.		
I eat an egg.		
I eat an avocado.		
I eat a tablespoon of extra virgin olive oil.		
I eat a palmful of nuts.		
I eat a handful of berries.		
I eat two handfuls of greens.		
I drink a supersmart smoothie.		
I eat a supersmart salad.		
I take supersmart supplements.		
I begin the day with a brainy breakfast.		
I avoid eating from the neurotoxin naughty list (page 75).		

I graze more, gorge less.		
I eat more nutrient-dense foods.		
I eat in omega balance.		
I eat mostly green-light, good-gut foods (page 89) to better my microbiome.		
I either don't drink alcohol, or I sip slowly with meals.		
MOVE SMART		
I move briskly an hour a day.		
I do isometrics ten minutes daily.		
I spend as many hours moving outdoors as I do sitting indoors.		
THINK SMART		
If I can't change it, I don't worry about it.		
I don't dwell on "dumb" decisions.		
I quickly trash toxic thoughts.		
I quickly repel ANTs.		
I "mind-mute" when needed.		
I do deep breathing when needed.		
I dwell on my "attitude of gratitude" several times a day.		
I do moving meditation.		
I enjoy mini-meditations several times a day.		
I start each day the peaceful way.		

I'm growing my God center.		
I frequently look at the "five things I like about me" on my cell phone.		
SLEEP SMART		
I unplug from artificial lights at least one hour before bedtime.		
I protect my sleep sanctuary.		
I drift off to sleep with an "attitude of gratitude."		
I wake up with an "attitude of gratitude."		

GRATITUDES

FROM DR. BILL:

Huge hugs of thanks to Lucille, my single mom, for teaching me a love of learning and channeling my "brainstorms" into the art of teaching.

A special thanks to my mentors who helped me to grow as a person and as a doctor, and to value the wonders and the health of the human brain:

- *Richard Van Praagh, MD*, professor of pediatrics at Harvard Medical School, who taught me the wisdom of how to study the complex brain: *show me the science.*
- *Dean Ornish, MD*, clinical professor of medicine, University of California, San Francisco, who helped me appreciate the concept of "doctor" first as preventer of disease and second as prescriber of medicines. Dr. Ornish's books have taught me to appreciate that brain health begins with vascular health.
- *Louis Ignarro, PhD*, for teaching me his Nobel Prize–winning research that inside our body is our own personal pharmacy, and how to protect it.
- *David Perlmutter, MD*, whose many books on the brain, and whose brain-health lectures that I have attended, emphasized that prevention is the key to brain health.

- *Max Lugavere,* author of *Genius Foods,* who, as a guest at our home for dinner, helped me appreciate the importance of getting the message of preventive brain health to more young adults.
- *David Katz, MD,* founding director of Yale University's Preventive Research Center, for teaching us doctors that brain health begins with lifestyle medicine.
- *Randy Hartnell,* owner of Vital Choice Seafood Company and my favorite fisherman, who taught me why seafood is the top nutrient for brain health.
- My coauthor, *Vincent Fortanasce, MD,* author of *The Anti-Alzheimer's Prescription,* clinical professor of neurology at the University of Southern California School of Medicine, and godfather to two of my and Martha's children, for the many hours we spent together studying how to convey preventive medicines for brain health in relatable and memorable terms.
- *Dave Asprey,* author of *Head Strong,* for the privilege of speaking on brain health at his conferences.
- *Ira Lott, MD,* former chief of neurology at University of California, Irvine, for giving me the privilege of teaching medical students preventive medicine for brain health.
- Vitamin D advocate *Carole Baggerly,* for teaching me how vitamin D deficiency is a contributor to Alzheimer's.

A huge brainful of thanks to *Bob and Dominique Hodgin,* directors of the Dr. Sears Wellness Institute, for taking my teachings on brain health to more than 10,000 certified health coaches in many countries throughout the world.

And a special thanks to *Andrew Newberg, MD,* neuroscientist, for his many writings on the relationship between a healthy spiritual life and a healthier mind.

FROM DR. VINCE:

First and foremost, I give thanks to my mother Rose and father Michael, and my wife Gayl.

To my sisters, Elaine and Joan, for their never-wavering support and friendship.

To my dear friends, Bill Sears, Bob Bancroft, and Phil Greene, and my colleagues, Drs. Lori Perino, Dino Claurizo, Mike Agron, Bill Preston, Bob Watkins Sr. and Jr., and Bill Caton, for your trust and support.

To my sons, Vinnie and Michael, and my granddaughter, Ava, for the joy they bring me.

To my daughter, Kaycee, who never gives up.

For all those who believe LIFE is an opportunity to love and do good.

FROM HAYDEN:

Thank you, Mom and Dad, for not only your amazing love and support, but for guiding me toward the tools to keep my brain healthy and thriving—and for creating a legacy that will help countless families for generations to come! Thank you also to my three children, whose unique brains inspire me to continue to learn and be curious.

FROM ALL THREE OF US:

This book could not have been written without the help of these special people:

- *Martha Sears, RN*, Dr. Bill's wife of fifty-three years. Thank you, Martha, for being our patient proofreader.
- *Gayl Fortanasce*, Dr. Vince's wife, whose profession as a psychotherapist was naturally reflected in this book.
- *Tracee Zeni*, Dr. Bill's diligent editorial assistant for more than twenty-eight years.
- *Matthew Sears*, our editing and research assistant, for his untiring search for the best science behind our transformation tools.
- *Jonathan Sears*, our online medical journal detective.
- Big smiles to our graphic artist, *Debbie Maze*, who makes readers smile with her illustrations.

- We deeply thank our literary agents, Denise Marcil and Anne Marie O'Farrell, at the Denise Marcil Agency, for making the perfect match with BenBella Books.
- A special thanks to the diligent staff at BenBella Books for their untiring patience and insightful suggestions: Leah Wilson, Editor-in-Chief; Jennifer Canzoneri, Marketing Director; Jessika Rieck, Deputy Production Manager; Adrienne Lang, Deputy Publisher; and Glenn Yeffeth, Publisher and CEO.

We thank you all for your contributions to making this book a good read!

SUGGESTED READING

OTHER RELATED BOOKS BY THE AUTHORS

Sears, William, and Erin Sears Basile. *The Dr. Sears T5 Wellness Plan: Transform Your Mind and Body, Five Changes in Five Weeks*. Dallas: BenBella Books, 2017.

Fortanasce, Vincent. *The Anti-Alzheimer's Prescription*. New York: Gotham Books, 2008.

ADDITIONAL SUGGESTED READING

Andreasen, Nancy C. *Brave New Brain: Conquering Mental Illness in the Era of the Genome*. New York: Oxford, 2001.

Asprey, Dave. *Game Changers: What Leaders, Innovators, and Mavericks Do to Win at Life*. New York: Harper Wave, 2018.

Borysenko, Joan. *The Plant Plus Diet Solution*. Carlsbad, CA: Hay House, 2014.

Cunnane, Stephen. *Survival of the Fattest*. Hackensack, NJ: World Scientific Publishing, 2005.

Doidge, Norman. *The Brain's Way of Healing: Remarkable Discoveries and Recoveries from the Frontiers of Neuroplasticity*. New York: Penguin Books, 2015.

Katz, David, and Stacey Coling. *Disease-Proof*. New York: Penguin Group, 2013.

Lewis, Michael. *When Brains Collide*. Austin: Lioncrest, 2016.

Lin, Steven. *The Dental Diet*. Carlsbad, CA: Hay House, 2018.

Lugavere, Max, and Paul Grewal. *Genius Foods*. New York: Harper Wave, 2018.

Masley, Steven, and Johnny Bowden. *Smart Fat*. New York: Harper One, 2016.

Newberg, Andrew, and Mark Robert Waldman. *Born to Believe: God, Science, and the Origin of Ordinary and Extraordinary Beliefs*. New York: Free Press, 2006.

O'Bryan, Tom. *You Can Fix Your Brain: Just 1 Hour a Week to the Best Memory, Productivity, and Sleep You've Ever Had*. New York: Rodale, 2018.

Ornish, Dean, and Anne Ornish. *Undo It! How Simple Lifestyle Changes Can Reverse Most Chronic Diseases*. New York: Ballantine Books, 2019.

Perlmutter, David, and Kristin Loberg. *Brain Maker*. New York: Little, Brown, 2015.

Ratey, John, and Eric Hagerman. *Spark*. New York: Little, Brown, 2008.

Sapolsky, Robert M. *Why Zebras Don't Get Ulcers: An Updated Guide to Stress, Stress-Related Diseases, and Coping*. New York: W. H. Freeman, 1994.

Selhub, Eva M., and Alan C. Logan. *Your Brain on Nature: The Science of Nature's Influence on Your Health, Happiness, and Vitality*. Ontario: Wiley Canada, 2012.

Siegel, Daniel J. *The Mindful Brain: Reflection and Attunement in the Cultivation of Well-Being*. New York: W. W. Norton, 2007.

Swaab, D. F. *We Are Our Brains: A Neurobiography of the Brain, from the Womb to Alzheimer's*. New York: Spiegel & Grau, 2014.

Wylde, Bryce. *Brainspan*. Landsdowne, Ontario: EnCompass Editions, 2020.

ENDNOTES AND REFERENCES

ENDNOTES

For a deeper understanding and scientific updates of some brain-health topics, visit DrSearsWellness.org/Healthy-Brain/ Endnotes.

PART I

Chapter 2: Eat Supersmart Foods: Medicines for Your Mind

Sheppard K, Cheatham C. Omega-6 to omega-3 fatty acid ratio and higher-order cognitive functions in 7- to 9-y-olds: a cross-sectional study. *Am J Clin Nutr*. 2013;98:659-667.

Qin B, et al. Fish intake is associated with slower cognitive decline in Chinese older adults. *J Nutr*. 2014;144(10):1579-1585.

Amminger G, et al. Longer-term outcome in the prevention of psychotic disorders by the Vienna omega-3 study. *Nature Communications*. 2015;6:7934.

Kiecolt-Glaser J, et al. Omega-3 supplementation lowers inflammation and anxiety in medical students: a randomized controlled trial. *Brain Behav Immun*. 2011;25(8):1725-1734.

Pottala J, et al. Higher RBC EPA + DHA corresponds with larger total brain and hippocampal volumes: WHIMS-MRI study. *Neurology*. 2014;82(5):435-442.

Witte V, et al. Long-chain omega-3 fatty acids improve brain function and structure in older adults. *Cerebral Cortex*. 2014;24:3059-3068.

Bloch M. Omega-3 fatty acid supplementation for the treatment of children with attention-deficit/hyperactivity disorder symptomatology: systematic review and meta-analysis. *J Am Acad Child Adolesc Psychiatr*. 2011;50(10):991-1000.

Devore E, et al. Dietary intake of berries and flavonoids in relation to cognitive decline. *Ann Neurol*. 2012;72(1):135-143.

Jakubowicz D, et al. High caloric intake at breakfast vs. dinner differentially influences weight loss of overweight and obese women. *Obesity*. 2013;21(12):2504-2512.

Martinez-Lapiscina E. Mediterranean diet improves cognition: the PREDIMED-NAVARRA randomised trial. *Neurol Neurosurg Psychiatr*. 2013;84(12):1318-1325.

Levine M, et al. Low protein intake is associated with a major reduction in IGF-1 cancer, and overall mortality in the 65 and younger but not older population. *Cell Metab*. 2014;19(3):407-417.

Institute of Preventive and Clinical Medicine. Advanced glycation end products and nutrition. *Physiol. Res*. 2002;51:313-316.

Vitali C, et al. HDL and cholesterol handling in the brain. *Cardiovasc Res*. 2014;103:405-413.

Kakehi E, et al. Non-diabetic glucose levels and cancer mortality: a literature review. *Curr Diabetes Rev*. 2018;14(5):434-445.

Avogaro A. Postprandial glucose: marker or risk factor? *Diabetes Care*. 2011;34(10):2333-2335.

Gerich J. Clinical significance, pathogenesis, and management of postprandial hyperglycemia. *Arch Intern Med*. 2003;163(11): 1306-1316.

Martins C, et al. Effects of restrained eating behaviour on insulin sensitivity in normal-weight individuals. *Physiol Behav*. 2009;96(4-5):703-708.

Sacks D. A1C versus glucose testing: a comparison. *Diabetes Care*. 201;34(2):518-523.

Onat A, et al. Fasting, non-fasting glucose and HDL dysfunction in risk of pre-diabetes, diabetes, and coronary disease in nondiabetic adults. *Acta Diabetologica*. 2013;50(4):519-528.

Ebenbichler C, et al. Postprandial state and atherosclerosis. *Curr Opin Lipidol*. 1995;6(5):286-290.

Sprague R, Ellsworth M. Vascular disease in pre-diabetes: new insights derived from systems biology. *Mo Med*. 2010;107(4):265-269.

Shimabukuro M, et al. Effects of dietary composition on postprandial endothelial function and adiponectin concentrations in healthy humans: a crossover controlled study. *Am J Clin Nutr*. 2007;86:923-928.

Kimattila S, et al. Chronic hyperglycemia impairs endothelial function and insulin sensitivity via different mechanisms in insulin-dependent diabetes mellitus. *Circulation*. 1996;94:1276-1282.

Jenkins D, et al. Rate of digestion of foods and postprandial glycaemia in normal and diabetic subjects. *Br Med J*. 1980;281(6232):14–17.

Suez J, et al. Artificial sweeteners induce glucose intolerance by altering the gut microbiota. *Nature*. 2014;514(7521):181-186.

Foster J, McVey Neufeld K. Gut-brain axis: how the microbiome influences anxiety and depression. *Trends Neurosci*. 2013;36(5): 305-312.

Mattson M, et al. Meal size and frequency affect neuronal plasticity and vulnerability to disease: cellular and molecular mechanisms. *J Neurochem*. 2003;84:417-431.

Cherbuin N, et al. Higher normal fasting plasma glucose is associated with hippocampal atrophy. *Neurology*. 2012;79(10):1019-1026.

Nitenberg A, et al. Postprandial endothelial dysfunction: role of glucose, lipids and insulin. *Diabetes Metab*. 2006;32 Spec No2:2S28-2S33.

Jankowiak J. Too much sugar may cause "brain decay." *Neurol.* 2004;63;E9-E10.

Jacka F, et al. Western Diet is associated with a smaller hippocampus: a longitudinal investigation. *BMC Med.* 2015;13:art no. 215.

Bjelland I, et al. Choline in anxiety and depression: the Hordaland Health Study. *Am J Clin Nutr.* 2009;90(4):1056-1060.

Ratliff J, et al. Consuming eggs for breakfast influences plasma glucose and ghrelin, while reducing energy intake during the next 24 hours in adult men. *Nutr Res.* 2010;30(2):96-103.

Wang D, et al. Specific dietary fats in relation to total and cause-specific mortality. *JAMA Intern. Med.* 2016;176(8):1134-1145.

Abuznait A, et al. Olive-oil-derived oleocanthal enhances B-amyloid clearance as a potential neuroprotective mechanism against Alzheimer's disease: in vitro and in vivo studies. *ACS Chem Neurosci.* 2013;4(6):973-982.

Weigle D, et al. A high-protein diet induces sustained reductions in appetite, ad libitum caloric intake, and body weight despite compensatory changes in diurnal plasma leptin and ghrelin concentrations. *Am J Clin Nutr.* 2005;82(1):41-48.

Galasso C, et al. On the neuroprotective role of astaxanthin: new perspectives? *Marine Drugs.* 2018;16(8).

Talbott S, et al. Astaxanthin supplementation reduces depression and fatigue in healthy subjects. *EC Nutr.* 2019;14.3:239-246.

Sears, William. *Natural Astaxanthin: Hawaii's Supernutrient.* 2015.

Sekikawa T, et al. Cognitive function improvement with astaxanthin intake: A randomized, double-blind, placebo-controlled study. *J Clin Biochem Nutr.* 2012;51(2):102-107.

Gupta V, et al. Garlic extract exhibits antiamyloidogenic activity on amyloid-beta fibrillogenesis: relevance to Alzheimer's disease. *Phytotherapy Research.* 2009;23(1):11-15.

Fernando W, et al. The role of dietary coconut for the prevention and treatment of Alzheimer's disease: mechanisms of action. *Br J Nutr.* 2015;114(1):1-14.

Kuriyama S, et al. Green tea consumption and cognitive function: a cross-sectional study from the Tsurugaya Project 1. *Am J Clin Nutr.* 2006;83:355-361.

Siri-tarino P, et al. Meta-analysis of protective cohort studies evaluating the associations of saturated fat with cardiovascular disease. *Am J Clin Nutr.* 2010;91(3):535-546.

Hu F, et al. Types of dietary fat and risk of coronary heart disease: a critical review. *J Am Coll Nutr.* 2001;20(1):5-19.

Devore E, et al. Dietary intake of berries and flavonoids in relation to cognitive decline. *Ann Neurol.* 2012;72(1):135-143.

Jakubowicz D, et al. High caloric intake at breakfast vs. dinner differently influences weight loss of overweight and obese women. *Obes (Silver Spring).* 2013;21(12):2504-2512.

Martinez-Lapiscina E, et al. Mediterranean diet improves cognition: the PREDIMED NAVARRA randomized trial. *J Neurol Neurosurg Psychiatr.* 2013;84(12):1318-1325.

Petrie M, et al. Beet root juice: An ergogenic aid for exercise and the aging bra. *J Gerontol A Biol Sci Med Sci.* 2017;72(9):1284-1289.

Crawford MA, et al. Evidence for the unique function of docosahexaenoic acid during the evolution of the modern hominid brain. *Lipids.* 1999;34:S39-S47.

Chapter 3: Be Good to Your Gut—Your Second Brain

Dinan T, et al. Psychobiotics: a novel class of psychotropic. *Biol Psychiatr.* 2013;74(10):720-726.

Cryan J, Dinan T. Mind-altering microorganisms: the impact of the gut microbiota on brain and behaviour. *Nat Rev Neurosci.* 2012;13(10):701-712.

Rafter J, et al. Dietary synbiotics reduce cancer risk factors in polypectomized and colon cancer patients. *Am J Clin Nutr.* 2007;85(2):488-96.

Collins S, et al. The interplay between the intestinal microbiota and the brain. *Nat Rev Microbiol.* 2012;10(11):735-742.

Furness J, et al. II. The intestine as a sensory organ: neural, endocrine, and responses. *Am J Physiol.* 1999;277(5):G922-G928.

Douglas-Escobar M, et al. Effect of intestinal microbial ecology on the developing brain. *JAMA Pediatr*. 2013;167(4):374-379.

Willeumier K, et al. Elevated BMI is associated with decreased blood flow in the prefrontal cortex using SPECT imaging in healthy adults. *Obes (Silver Spring)*. 2011;19(5):1095-1097.

Agusti A, et al. Interplay between the gut-brain axis, obesity and cognitive function. *Frontiers Neurosci*. 2018;12:155.

Calvani R, et al. Of microbes and minds: a narrative review on the second brain aging. *Frontiers Med*. 2018;5:53.

Cani P, Knauf C. How gut microbes talk to organs: the role of endocrine and nervous routes. *Mol Metab*. 2016;5(9):743-752.

Kelly J, et al. Transferring the blues: Depression-associated gut microbiota induces neurobehavioural changes in the rat. *J Psychiatr Res*. 2016;82:109-118.

O'Sullivan O, et al. Exercise and the microbiota. *Gut Microbes*. 2015;6(2):131-136.

Chapter 4: Move Smart!

Gomes da Silva S, et al. Early exercise promotes positive hippocampal plasticity and improves spatial memory in the adult life of rats. *Hippocampus*. 2012;22(2):347-358.

Ahmadi N, et al. Long-term regular exercise promotes memory and learning in young but not in older rats [published online ahead of print December 3, 2007]. *Pathophysiology*. 2008;15(1):9-12.

Schneider S, et al. Brain and exercise: a first approach using electrotomography. *Med Sci Sports Exerc*. 2010;42(3):600-607.

Colombe S, et al. Aerobic exercise training increases brain volume in aging humans. *J Gerontol Ser A*. 2006;61(11):1166-1170.

Dietrich A, McDaniel W. Endocannabinoids and exercise. *Br J Sports Med*. 2004;38(5):536-541.

Emery C, et al. Exercise accelerates wound healing among healthy older adults: a preliminary investigation. *J Gerontol Ser A*. 2005;60(11):1432-1436.

Best J, et al. Long-term effects of resistance exercise training on cognition and brain volume in older women: results from a randomized controlled trial. *J Int Neuropsychol Soc.* 2015;21(10):745-756.

McDonnell M, et al. A single bout of aerobic exercise promotes motor cortical neuroplasticity. *J Appl Physiol.* 2013;144(9):1174-1182.

Ding Q, et al. Insulin-like growth factor 1 interfaces with brain-derived neurotrophic factor-mediated synaptic plasticity to modulate aspects of exercise-induced cognitive function. *Neurosci.* 2006;140(3):823-833.

Gomez-Pinilla F, et al. BDNF and learning: evidence that training promotes learning within the spinal cord by up-regulating BDNF expression. *Neurosci.* 2007;148(4):893-906.

Taylor A, Sullivan W. Coping with ADD. The surprising connection to green play settings. *Environ Behav.* 2001:33(1):54-77.

Ploughman M. Exercise is brain food: the effects of physical activity on cognitive function. *Dev Neurorehabil.* 2008;11(3):236-240.

Viadero D. Exercise seen as priming pump for students' academic strides. *Educ Week.* 2008;27(23):14-15.

Brinke L, et al. Aerobic exercise increases hippocampal volume in older women with probable mild cognitive impairment: a 6-month randomized controlled trial. *Br J Sports Med.* 2015;49(4):248-254.

Maddock R, et al. Acute modulation of cortical glutamate and GABA content by physical activity. *J Neurosci.* 2016;36(8):2449-2457.

Sterling PB, et al. Exercise activates the endocannabinoid system. *Cogn Neurosci.* 2003;14(17):2209-2211.

Dietrich A, McDaniel W. Endocannabinoids and exercise. *Br J Sports Med.* 2004;38(5):536-541.

Deslandes A, et al. Exercise and mental health: many reasons to move. *Neuropsychobiol.* 2009;59(4):191-198.

Wrann C, et al. Exercise induces hippocampal BDNF through a PGC-1a/FNDC5 pathway. *Cell Metab.* 2013;18(5):649-659.

Chaddock L, et al. A neuroimaging investigation of the association between aerobic fitness, hippocampal volume, and memory performance in preadolescent children. *Brain Res.* 2010; 1358:172-183.

Chaddock L, et al. A functional MRI investigation of the association between childhood aerobic fitness and neurocognitive control. *Biol Psychol*. 2012;89(1):260-268.

Kramer A, et al. Ageing, fitness and neurocognitive function. *Nature*. 1999;400(6743):418-419.

Lucertini F, et al. High cardiorespiratory fitness is negatively associated with daily cortisol output in healthy aging men. *PLoS One*. 2015;10(11):e0141970.

Mattson M. Energy intake and exercise as determinants of brain health and vulnerability to injury and disease. *Cell Metab*. 2012;16(6):706-722.

Li Q, et al. Effect of phytonide from trees on human natural killer cell function. *Int J Immunopathol*. 2009;22(4):951-959.

Wahl P. Hormonal and metabolic responses to high intensity interval training. *J Sports Med Doping Stud*. 2013;3:e132.

Gillen J, et al. Acute high-intensity interval exercise reduces the postprandial glucose response and prevalence of hyperglycaemia in patients with type 2 diabetes. *Diabetes Obes Metab*. 2012;14(6):575-577.

Boecker H, et al. The runner's high: opioidergic mechanisms in the human brain. *Cerebral Cortex*. 2008;18(11):2523-2531.

Morita E, et al. Psychological effects of forest environments on healthy adults: shinrin-yoku (forest-air bathing, walking) as a possible method of stress reduction. *Public Health*. 2007;121(1):54-63.

Harvey S, et al. Exercise and the prevention of depression: results of the HUNT cohort study. *Am J Psychiatr*. 2018;175(1):28-36.

Dimeo F, et al. Benefits from aerobic exercise in patients with major depression: a pilot study. *Br J Sports Med*. 2001;35(2):114-117.

Stubbs J, et al. A decrease in physical activity affects appetite, energy, and nutrient balance in lean men feeding ad libitum. *Am J Clin Nutr*. 2004;79(1):62-69.

Kohl H III, Cook H., eds. Physical activity, fitness, and physical education: effects on academic performance. In: *Educating the Student Body: Taking Physical Activity and Physical Education to School*.

Washington, DC: National Academies Press, 2013. www.nap.edu/read/18314/chapter/6.

Martinsen E, et al. Effects of aerobic exercise on depression: a controlled study. *Br Med J (Clin Res Ed)*. 1985;291(6488):109.

Kavouras S, et al. Physical activity and adherence to Mediterranean diet increase total antioxidant capacity: the ATTICA study. *Cardiol Res Pract*. 2010;2011:248626.

Gomez-Pinilla F, et al. Voluntary exercise induces a BDNF-mediated mechanism that promotes neuroplasticity. *J Neurophysiol*. 2002;88(5):2187-2195.

Chapter 5: Think Smart: Eight Stressbusters to Think-Change Your Brain

Erickson K, et al. Exercise training increases size of hippocampus and improves memory. *PNAS*. 2011;108(7):3017-3022.

Gerberg PL, et al. The effect of breathing, movement, and meditation on psychological and physical symptoms and inflammatory biomarkers in inflammatory bowel disease: a randomized controlled trial. *Inflamm Bowel Dis*. 2015;21(12):2886-2896.

Epel E, et al. Accelerated telomere shortening in response to life stress. *Proc Natl Acad Sci USA*. 2004;101(49):17312-17315.

Miller M, et al. Impact of cinematic viewing on endothelial function. *Heart*. 2006;92:261-262.

Tang Yi-Yuan, et al. Mechanisms of white matter changes induced by meditation. *Proc Natl Acad Sci USA*. 2012;109(26):10570-10574.

Hasenkamp W, et al. Mind wandering and attention during focused meditation: a fine-grained temporal analysis of fluctuating cognitive states. *Neuroimage*. 2012;59(1):750-760.

Lazar S, et al. Meditation experience is associated with increased cortical thickness. *Neuroreport*. 2005;16(17):1893-1897.

Luders E, et al. Enhanced brain connectivity in long-term meditation practitioners. *Neuroimage*. 2011;57(4):1308-1316.

Brown WR, et al. Review: cerebral microvascular pathology in aging and neurodegeneration. *Neuropathol Appl Neurobiol*. 2011;37:56-74.

Colombe SJ, et al. Aerobic exercise training increases brain volume in aging humans. *J Geron A Biol Sci Med Sci*. 2006;61:1166-1170.

Kaptchuk T, et al. Placebo effects in medicine. *N Engl J Med*. 2015;373:8-9.

Chapter 6: Sleep Smart: How to Get a Good Night's Sleep

Glaser R. Stress-associated immune dysregulation and its importance for human health: a personal history of psychoneuroimmunology. *Brain Behav Immun*. 2005;19(1):3-11.

Xie L, et al. Sleep drives metabolite clearance from the adult brain. *Science*. 2013;342(6156):373-377.

Lee H, et al. The effect of body posture on brain glymphatic transport. *J Neurosci*. 2015;35(31):11034-11044.

Breus M. *Good Night*. New York: Penguin, 2006.

Walker M. *Why We Sleep*. New York: Scribner, 2017.

PART II

Chapter 7: How to Heal from Anxiety and Depression

Kramer A, Erickson K. Effects of physical activity on cognition, well-being, and brain: human interventions. *Alzheimer's Dement*. 2007;3(2 Suppl):S45-S51.

Zwilling C, et al. Nutrient biomarker patterns, cognitive function, and fMRI measures of network efficiency in the aging brain. *Neuroimage*. 2019;188:239-251.

Zamroziewicz M, Barbey A. Nutritional cognitive neuroscience: innovations for healthy brain. *Front Neurosci*. 2016;10:240.

Lieverse R, et al. Bright light treatment in elderly patients with nonseasonal major depressive disorder: a randomized placebo-controlled trial. *Arch Gen Psychiatr*. 2011;68(1):61-70.

Gangwisch J, et al. High glycemic index diet as a risk factor for depression: analyses from the women's health initiative. *Am J Clin Nutr*. 2015;102(2):454-463.

Soczynska J, et al. Mood disorders and obesity: understanding inflammation as a pathophysiological nexus. *Neuromol Med*. 2010;13(2):93-116.

Jacka F, et al. Association of western and traditional diets with depression and anxiety in women. *Am J Psychiatr.* 2010;167(3):305-311.

Jacka F, et al. A randomised controlled trial of dietary improvement for adults with major depression (the "SMILES" trial). *BMJ Med.* 2017;15(1):23.

Chapter 8: Preventing and Reversing Alzheimer's Disease and Other Neurodegenerative Diseases

Feng C, et al. Hyperhomocysteinemia associates with small vessel disease more closely than large vessel disease. *Int J Med Sci.* 2013;10:408-412.

Bredesen D. Reversal of cognitive decline: a novel therapeutic program. *Aging (Albany NY).* 2014;6(9):707-717.

Ferrucci L, et al. Serum IL-6 level and the development of disability in older persons. *J Am Geriatr Soc.* 1999;47(6):639-646.

Franceschi C, et al. Genes involved in immune response/inflammation, IGF1/insulin pathway and response to oxidative stress play a major role in the genetics of human longevity: the lesson of centenarians. *Mech Ageing Dev.* 2005;126(2):351-361.

Ngandu T, et al. A 2 year multidomain intervention of diet, exercise, cognitive training, and vascular risk monitoring versus control to prevent cognitive decline in at-risk elderly people (FINGER): a randomized controlled trial. *Lancet.* 2015;385(9984):2255-2263.

Head D, et al. Exercise engagement as a moderator of APOE effects on amyloid deposition. *Arch Neurol.* 2012;69(5):636-643.

Morris M, et al. MIND diet slows cognitive decline with aging. *Alzheimer's Dement.* 2015;11(9):1015-1022.

Willette A, et al. Association of insulin resistance with cerebral glucose uptake in late middle-aged adults at risk for Alzheimer's disease. *JAMA Neurol.* 2015;72(9):1013-1020.

Whitmer R, et al. Central obesity and increased risk of dementia more than three decades later. *Neurology.* 2008;71(14):1057-1064.

Debette S, et al. Visceral fat is associated with lower brain volume in healthy middle-aged adults. *Ann Neurol.* 2010;68(2):136-144.

Ho A, et al. Obesity is linked with lower brain volume in 700 AD and MCI patients. *Neurobiol Aging*. 2010;31(8):1326-1339.

Brown W, Thore C. Review: cerebral microvascular pathology in aging and neurodegeneration. *Neuropathol Appl Neurobiol*. 2011;37(1):56-74.

Cassilhas R, et al. The impact of resistance exercise on the cognitive function of the elderly. *Med Sci Sports Exerc*. 2007;39(8):1401-1407.

Schuster L, et al. Normal aging and imaging correlations. *Radiologe*. 2011;51(4):266-272.

Chapter 9: ADHD: Celebrating the "Difference," Healing the "Disorder"

Chevrin R, et al. Symptoms of sleep disorders, inattention, and hyperactivity in children. *Sleep*. 1997;20(12):1185-1192.

Waschbusch D, et al. Are there placebo effects in the medication treatment of children with attention-deficit hyperactivity disorder? *J Dev Behav Pediatr*. 2009;30(2):158-168.

Joseph N, et al. Oxidative stress and ADHD: A meta-analysis. *J Atten. Disord*. 2015;19(11):915-924.

Warner-Schmidt J, Duman R. Hippocampal neurogenesis: opposing effects of stress and antidepressant treatment. *Hippocampus*. 2006;16(3):239-249.

Millichap J, Yee M. The diet factor in attention-deficit/hyperactivity disorder. *Pediatrics*. 2012;129(2):330-337.

Taylor A, et al. Coping with ADD: The surprising connection to green play settings. *Environ Behav*. 2001;33(1):54-77.

Sears W, Thompson L. *The A.D.D. Book*. New York: Little, Brown, 1998.

Chapter 10: Drink Smart: Preventing Alcohol Disease

GBD 2016 Alcohol Collaborators. Alcohol use and burden for 195 countries and territories, 1990–2016: a systematic analysis for the Global Burden of Disease Study 2016. *Lancet*. 2018;392(10152):1015-1035.

Topiwala A, et al. Moderate alcohol consumption as risk factor for adverse brain outcomes and cognitive decline: longitudinal cohort study. *BMJ*. 2017;357:j2353.

Pawlosky R, Salem N Jr. Ethanol exposure causes a decrease in docosahexaenoic acid and an increase in docosapentaenoic acid in feline brains and retinas. *Am J Clin Nutr*. 1995;61:1284-1289.

Researchers point to air pollution as cause of magnetite nanoparticles in brain. Alzheimer's Research UK website. www.alzheimersresearchuk .org/researchers-point-air-pollution-cause-magnetite-nanoparticles -brain/. Published September 5, 2016.

Purohit V, et al. Alcohol, intestinal bacterial growth, intestinal permeability to endotoxins and medical consequences. *Alcohol*. 2008;42(5):349-361.

Part III: More Questions You May Have About Brain Health

Weiss B. Vulnerability of children and the developing brain to neurotoxic hazards. *Environmental Health Perspectives*. 2000;108(Suppl 3):375-381.

Stauber J, Florence T. A comparative study of copper, lead, cadmium, and zinc in human sweat and blood. *Science Total Environment*. 1988;74:235-247.

Johansson O. Disturbance of the immune system by electromagnetic fields—a potentially underlying cause for cellular damage and tissue repair reduction which could lead to disease and impairment. *Pathophysiology*. 2009;16(2-3):157-177.

Bert P, et al. The effects of air pollution on the brain: a review of studies interfacing environmental epidemiology and neuroimaging. *Curr Environ Health Rep*. 2018;5(3):351-364.

INDEX

ABOUT THE AUTHORS

William Sears, MD, has been advising busy parents on how to raise healthier families for more than forty years, and now turns his attention to the specialty of lifestyle medicine. He is the cofounder of AskDrSears.com and the Dr. Sears Wellness Institute, which has certified more than 10,000 Health Coaches around the world. He has served as a voluntary professor at the University of Toronto, University of South Carolina, University of Southern California School of Medicine, and University of California, Irvine.

As a father of eight children, he coached Little League sports for twenty years, and together with his wife Martha has written more than forty books and countless articles on parenting, nutrition, and healthy aging. He serves as a consultant for TV, magazines, radio, and other media, and his website AskDrSears.com is one of the most popular health and parenting sites. Dr. Sears and his contribution to family health were featured on the cover of *TIME* magazine in May 2012. He is noted for his science-made-simple-and-fun approach to family health.

Vincent M. Fortanasce, MD, is ranked as one of the best medical specialists in North America and has treated such high-profile individuals as Pope John Paul II and Major League Baseball Hall of Famer Tommy Lasorda. For nearly four decades, Dr. Fortanasce has helped thousands of people as a world-renowned neurologist and rehabilitation specialist. He has appeared as a medical expert on *60 Minutes*, *Today*, *Dr. Phil*, *Dateline*, CNN's *Paula Zahn Now*, *Hard Ball with Chris Mathews*, XM satellite radio, and many more national and local television and radio shows. Dr. Fortanasce is a regular spokesperson for the California Medical Association at the senate and legislature assemblies and has been quoted in the *New York Times*, *Sports Illustrated*, *USA Today*, *U.S. News & World Report*, *Time* magazine, and many other prestigious publications. He also hosts his own syndicated radio program, St. Joseph's Radio Presents.

~~~~~~~~~~~~~~~~~~~~~~~~

Hayden Sears, MA, mother of three, is a certified health and nutrition coach who loves helping families and individuals on their journey toward better health. The oldest daughter of Dr. William and Martha Sears, she has worked with the Sears Family Pediatrics medical practice for more than fifteen years as Wellness Coordinator. She also contributes to the content of AskDrSears.com; has been a guest on TV shows and news stations sharing nutrition tips, healthy meal options, and the benefits of baby-wearing; and co-hosts the *Dr. Sears Family Podcast*. Hayden owns a Juice Plus+ virtual franchise and travels all over the world speaking about how to keep families healthy. She received her MA from Azusa Pacific University and resides in Southern California.